Mathematical Representation at the Interface of Body and Culture

A volume in
International Perspectives on Mathematics Education—
Cognition, Equity, & Society

Series Editors:
Bharath Sriraman, *The University of Montana*
Lyn English, *Queensland University of Technology*

International Perspectives on Mathematics Education—Cognition, Equity, and Society

Bharath Sriraman and Lyn English, Series Editors

Mathematical Representations at the Interface of Body and Culture (2009)
edited by Wolff-Michael Roth

Challenging Perspectives on Mathematics Classroom Communication (2006)
edited by Anna Chronaki and Iben Maj Christiansen

Which Way Social Justice in Mathematics Education (2005)
edited by Leone Burton

Mathematics Education Within the Postmodern (2004)
edited by Margaret Walshaw

Research Mathematics Classrooms:
A Critical Examination of Methodology (2002)
edited by Simon Goodchild and Lyn English

Mathematical Representation at the Interface of Body and Culture

edited by

Wolff-Michael Roth
University of Victoria, Canada

Information Age Publishing, Inc.
Charlotte, North Carolina • www.infoagepub.com

Library of Congress Cataloging-in-Publication Data

Mathematical representation at the interface of body and culture / edited by Wolff-Michael Roth.
  p. cm. -- (International perspectives on mathematics education, cognition, equity & society)
 Includes bibliographical references.
 ISBN 978-1-60752-130-3 (pbk.) -- ISBN 978-1-60752-131-0 (hardcover) 1. Mathematics--Study and teaching--Social aspects. I. Roth, Wolff-Michael, 1953-
 QA11.2.M2773 2009
 510.71--dc22

                                                                2009014762

CONTENTS

Series Preface *vii*

Preface *xi*

1. Social Bodies and Mathematical Cognition: An Introduction
   Wolff-Michael Roth *1*

PART A: MOVING AND TRANSFORMING BODIES IN/AS MATHEMATICAL PRACTICE

Editor's Section Introduction *19*

2. Transformation Geometry from an Embodied Perspective
   Laurie D. Edwards *27*

3. Signifying Relative Motion: Time, Space and the
   Semiotics of Cartesian Graphs
   Luis Radford *45*

4. What Makes a Cube a Cube? Contingency in Abstract,
   Concrete, Cultural and Bodily Mathematical Knowings
   Jean-François Maheux, Jennifer S. Thom, and
   Wolff-Michael Roth *71*

5. Embodied Mathematical Communication and the
   Visibility of Graphical Features
   Wolff-Michael Roth *95*

Editor's Section Commentary *123*

PART B: EMERGENCE OF OBJECTS AND UNDERSTANDING

Editor's Section Introduction *131*

6. Supporting Students' Learning About Data Creation
Paul Cobb and Carrie Tzou *135*

7. How Do You Know Which Way the Arrows Go?
The Emergence and Brokering of a Classroom Math Practice
Chris Rasmussen, Michelle Zadieh, and Megan Wawro *171*

8. Inscription, Narration and Diagram-Based Argumentation:
Narrative Accounting Practices in Primary Mathematics Classes
Götz Krummheuer *219*

Editor's Section Commentary *245*

PART C: STEPS TOWARD RETHINKING
MATHEMATICS EDUCATION

Editor's Section Introduction *251*

9. And so ...?
Brent Davis *257*

10. Expressiveness and Mathematics Learning
Ian Whitacre, Charles Hohensee, and Ricardo Nemirovsky *275*

11. Gesture, Abstraction, and the Embodied Nature of Mathematics
Rafael E. Núñez *309*

Editor's Section Commentary *329*

PART D: EPILOGUE

12. Appreciating the Embodied Social Nature of
Mathematical Cognition
Wolff-Michael Roth *335*

About the Authors *351*

SERIES PREFACE

Wolff-Michael Roth's *Mathematical Representation at the Interface of Body and Culture* signals the revival of the series International Perspectives on Mathematics Education—Cognition, Equity, and Society, after a three-year hiatus. The series was in grave danger of going extinct due to the unfortunate passing away of its founder Leone Burton. Information Age Publishing entrusted us the task of keeping this series alive and to continue to give voice to prescient topics of importance to the international mathematics education community. We are pleased and privileged to include Roth's edited compilation on *Mathematical Representation at the Interface of Body and Culture* for many reasons.

Mathematics education has undergone numerous paradigm shifts, such as theory and methodology borrowing from psychology, sociology, anthropology, among others, infatuation with numerous -isms and a continued identity crisis. A critical question, however, that has been posed by scholars now and in previous decades is whether our paradigm shifts are genuine. That is, are we replacing one particular theoretical perspective with another that is more valid or more sophisticated for addressing the hardcore issues we confront or, as Alexander and Winne (2006) ask, is it more the case that theoretical perspectives move in and out of favor as they go through various transformations and updates? If so, is it the voice that speaks the loudest that gets heard? Who gets suppressed? The rise of constructivism in its various forms is an example of a paradigm that appeared to drown out many other theoretical voices during the 1990s (Goldin, 2003). Embodied mathematics made its appearance with the work of Lakoff and Núñez (2000), yet the bold ideas proposed in *Where Does Mathematics Come From*, received very little attention from mathematics education researchers in terms of systemic follow ups in teaching, learning and researching.

Similarly, even though Lev Vygotsky's work is found cited in the massive amount of literature in mathematics education that use social constructivist frameworks, very little attention is paid to his cultural-historical activity theory, which has simultaneous orientation with embodied operations and the social dimensions allowing for a theorization of the intricate relationships between individual and social cognition (Roth, 2007). In essence, the question we need to consider is whether we are advancing professionally in our theory development. Paradigms, such as constructivism, which became fashionable in mathematics education over recent decades, tended to dismiss or deny the integrity of fundamental aspects of mathematical and scientific knowledge. We agree with Goldin (2003) that "It is time to abandon, knowledgeably and thoughtfully, the dismissive fads and fashions—the 'isms'—in favor of a unifying, non-ideological, scientific and eclectic approach to research, an approach that allows for the *consilience* of knowledge across the disciplines" (p. 176)

Five years ago, Burton (2004) proposed an epistemological model of "coming to know mathematics" consisting of five interconnecting categories, namely the person and the social/cultural system, aesthetics, intuition/insight, multiple approaches, and connections, grounded in the extensive literature base of mathematics education, sociology of knowledge and feminist science, in order to address the challenges of objectivity, homogeneity, impersonality, and incoherence. Burton proposed that we view mathematics as a socio-cultural artifact, part of a larger cultural system as opposed to the Platonist objective view, and in order to substantiate her epistemological model, the work of Lakoff and Núñez (2000) on embodiment and Rotman (2000) on semiotics was extensively appealed to. *Mathematical Representation at the Interface of Body and Culture* presents a convergence of numerous ideas that have intersected with mathematics education but not been properly followed up in terms of their significance for the field. The present book fills a major void in our field by giving a masterfully edited coherent synthesis of the ongoing work on embodiment and representations in mathematics, grounded in cultural-historical activity theory and thus presents a strong case that much progress can and has been made in mathematics education, and indeed much can be learned from the work of the authors carefully cast in the book by Wolff-Michael Roth.

Bharath Sriraman & Lyn English
Series Editors

REFERENCES

Alexander, P. A., & Winne, P.H. (Eds.). (2006). *Handbook of educational psychology* (2nd ed.). Mahwah, NJ: Lawrence Erlbaum Associates.

Burton, L. (2004). *Mathematicians as enquirers: Learning about learning mathematics*. Dordrecht, The Netherlands: Kluwer Academic Publishers.

Goldin, G. A. (2003). Developing complex understandings: On the relation of mathematics education research to mathematics. *Educational Studies in Mathematics, 54*, 171-202.

Lakoff, G., & Núñez, R. (2000). *Where mathematics comes from*. New York: Basic Books.

Roth, W.-M. (2007). Emotions at work: A contribution to third-generation cultural historical activity theory. *Mind, Culture and Activity, 14*, 40–63.

Rotman, B. (2000). *Mathematics as sign: Writing, imagining, counting*. Stanford, CA: Stanford University Press.

PREFACE

The purpose of this book is to bring together two bodies of literature that are often treated as different approaches to thinking about mathematical knowing and learning. The first body of literature articulates the intricate relationship between the individual and the social configuration of which she or he is part. That is, this body of work, running alongside more traditional constructivist and psychological approaches, shows that what happens at the collective level in a classroom both constrains and affords opportunities for what individuals do (their practices). A second major shift in mathematical theorizing occurred during the past decade: There is now an increasing focus on the embodied and bodily manifestation of mathematical knowing even though its practitioners have come from quite different literatures and backgrounds.

This edited book situates itself at the intersection of these theoretical and focal concerns. All chapters deal with mathematical inscriptions as these are interpreted, produced, or both. The domains include geometry (Edwards, Maheux et al.), data-representing inscriptions including line graphs, histograms, and scatter plots (Cobb, Krummheuer, Nemirovsky, Radford, Roth), and functions, equations and multiplication (Davis, Núñez, Rasmussen et al.). In each case people collectives are involved, sometimes in whole-class settings (Cobb, Davis, Edwards, Maheux et al., Krummheuer, Rasmussen), at other times in two- or three-person transactions (Nemirovsky, Radford, Roth); multiple people are involved even in the presentation of a mathematician (Núñez), who is not just talking to and for himself but inherently is oriented to another (researcher, person videotaping). In all chapters, the current culture both at the classroom and at the societal level comes to be expressed and provides opportunities for expressing oneself in particular ways; and these expressions always are bodily expressions of body-minds (especially emphasized in Edwards, Maheux et al., Nemirovsky, Núñez, Radford, Roth). As a collective, the chapters focus on mathematical knowledge as an aspect or attribute of mathematical performance; that is, mathematical knowing is in the doing. Theorizing this doing appears to be more fruitful for subsequent implications to teaching than theorizing what happens in terms of mental substrates structured in particular ways as conceived by those who draw on conceptual change perspectives or on traditional cognitive psychological concepts. My own ethnomethodologically and conversational analytically informed study (chapter 5) brings out the simultaneously social and bodily embodied nature of mathematical cognition in situations that are intelligible to, and intelligibly created by, the co-participants to conversational transaction.

The collection of chapters as a whole articulates for readers the important aspects of mathematical cognition that are produced and observable at the interface between the body (both human and those of [inherently material] inscriptions) and culture. The book is organized into three sections: (a) moving and transforming bodies in/as mathematical practice, (b) emergence of objects and understanding, and (c) steps toward the re-thinking of mathematics education. To facilitate an integrative reading across chapters, I introduce and conclude each of the three sections with commentaries designed to highlight similarities and differences between chapters and to integrate them into the single narrative that ties the book together into a coherent whole. Cultural-historical activity theory, with its focus on the material and ideal moments of praxis—enhanced by phenomenological perspectives that focus on first person experiences of praxis—is used as a framework for integrating the chapters into a single overarching argument.

Victoria, British Columbia
February 2009

CHAPTER 1

SOCIAL BODIES AND MATHEMATICAL COGNITION

An Introduction

Wolff-Michael Roth

In this book, two bodies of literature come together: mathematical knowing in communities of practice and the embodied nature of mathematical knowing. The purpose of this introductory chapter is to (a) set an ever-so-brief historical context of developments in the late twentieth-century; (b) provide a philosophical context for the debate about the nature of mathematical knowing as idealist or embodied; and (c) articulate cultural-historical activity theory as dialectical materialist alternative that can be used to integrate into a single framework the two literatures emphasizing the embodied and social nature of mathematical knowing.

LATE TWENTIETH-CENTURY DEVELOPMENTS

Over the past two decades, the theoretical interests of mathematics educators have changed substantially—as any brief look at the titles and abstracts of articles in relevant journals shows. In the 1980s, mathematical knowing and learning were explained in psychological models, focusing on motivation, interest, abilities, information processing, and conceptual (including alternative) frameworks. Jean Piaget's concepts of accommodation and assimilation reigned as a pair of key theoretical concepts that many mathematics educators have drawn on to explain observations made in individual interviews, teaching experiments, and quantitative classroom studies. His work alerted mathematics educators to the role of bodily experiences in the elaboration of mind especially drawing also on the concept of reflective abstraction. Piaget's perspective, however, is somewhat limited because he operated from Kantian principles attempting to identify formal knowledge that exists in excess of practical knowledge of the world without providing a framework how this might go. Moreover, Piaget did not address a well-known learning paradox, which questions the possibility of "constructing" abstract (formal) representations and forms of reasoning when the tools, objects, and material ground on and with which this construction occurs all are of lower complexity (e.g., Bereiter, 1985). He further did not address the question of how students can *intend* to learn something when they do not yet have the knowledge required to formulate the object of this intention. How can anyone who stands in a clearing (knowing) intend to learn the unseen in the surrounding dark without already knowing it?

Social approaches have the potential to overcome the two limitations articulated in the learning paradox, because more complex structures and forms of reasoning are already available in the collective (culture, society), where students can appropriate them in and through participation. The Russian psychologist Lev Vygotsky (1978) holds that any higher-order psychological function exists in social and societal relations, where individuals appropriate this function by participating in the relation. From an a-posteriori perspective, one can now detect early signs of a way of thinking that has revolutionized mathematics education since. Early papers of Paul Cobb (e.g., 1985) show that whereas he attended to individual cognition, the learner's motivation, and the (radical) construction of individual worlds, a beginning attention to the social aspects of mathematical cognition and an awareness of the literature in the social studies of science (e.g., Karin Knorr-Cetina, Barry Barnes, Thomas Kuhn) was there as well.

Largely through the work of Paul Cobb and his various collaborators, mathematics educators have come to be attuned to the intricate relationship between the individual learner of mathematics and the social configuration of which s/he is a constitutive part. That is, this body of work, running alongside more traditional constructivist and psychological approaches, shows that what happens at the collective level in a classroom both constrains and affords opportunities for what individuals do mathematically (their practices). Increasingly, researchers focused on the mediational role of socio-mathematical norms and how these emerged from the enacted lessons. This perspective often is articulated in terms of the relations between the inter- and intra-mental (or psychological) processes that Lev Vygotsky (1978) has articulated. Hereby, practices and processes in which a learner participates with others (inter-psychological plane) emerge and then operate on their own on an intra-psychological plane. It is apparent that activity *leads* development, which means, that learners participate in relations where mathematical activity is done *prior to* or *at the leading edge of* their own developmental capacity to do the mathematical task on their own.

A second major shift in mathematical theorizing has occurred in the course of the past decade: there is an increasing focus on the embodied and bodily manifestations of mathematical knowing (e.g., Lakoff & Núñez, 2000). Mathematics educators now working from this perspective have come to their position from quite different bodies of literatures: For some, linguistic concerns and mathematics as material praxis lay at the origin for their concerns (e.g., George Lakoff, Rafael Núñez); others came to their position through the literature on the situated nature of cognition (e.g., Jean Lave); and yet another line of thinking emerged from the work on embodiment that Humberto Maturana and Francisco Varela advanced in the late 1970s and early 1980s. Whatever the historical origins of their thinking, mathematics educators taking an embodiment perspective presuppose that it is of little use to think of mathematical knowing in terms of transcendental concepts somehow recorded in the brain, but rather, that we need to think about conceptual knowing as mediated by the human body, which, because of its *senses*, is the very condition and origin of sense. This sense frequently is used in metaphorical extensions to understand other domains often inaccessible or inexpressible in everyday experiences. The problems with the assumption of a transcendental nature of mathematical concepts is well captured in the "grounding problem" that artificial intelligence researchers have identified: How can mental concepts refer to anything in the social and material world, that is, how can they be grounded? We already see in the preceding paragraphs that Vygotsky and his followers would say that the grounding problem is an artifact of theory and method. His own approach—whereby participation in the social and material world (very much an embodiment approach) is the very condition for anything mental to occur—constitutes the solution to the grounding problem.

In this book, as part of my introduction and commentary to each of the three sections, I use cultural-historical activity theory, an outcrop of Vygotsky's way of thinking about the development of human consciousness, because it combines both the embodied and the social moments of mathematical cognition. Because embodiment and cultural-historical activity theoretic approaches have emerged as a reaction to idealist approaches to theorizing human cognition, I begin by providing an account of the idealist perspective on the mind that has its origin in the work of Immanuel Kant and, following, him, in Jean Piaget and (radical) constructivism. I then present just those aspects of dialectical materialist cultural-historical activity theory that I subsequently mobilize to contextualize the different contributions.

MATHEMATICAL KNOWING: PHILOSOPHICAL AND EPISTEMOLOGICAL CONSIDERATIONS

In mathematical cognition as it was presented by Immanuel Kant, I follow the general usage of the terms *concepts* and *conceptions* (also *conceptualizations*), using the former for the shared public aspects of some idea—the concept of triangle as a geometrical figure that can be constructed by three non-parallel lines on a plane—and the conception of triangle, which is the way students and others think about triangles. Geometry exists both virtually, as in the written texts that make the subject available independent of oral tradition, and real, in that it can be practiced and thereby brought to life at any time, in public, open to scrutiny (Husserl, 1939). The transcendental nature of geometry—though the very origin of the first geometrical (meaningful, sense-giving) act remains in darkness— derives precisely from the fact that it can be reproduced and transformed at any point and at any place in precisely the same manner. It exists in and through this possibility to enact and re-enact (re-activate) geometrical concepts, which does not exist other than in sedimented and crystallized form. This usage provides continuity with respect to the theoretical concept of misconception that some mathematics educators use as framework for studying mathematical cognition generally and mathematical cognition in geometry more specifically; it also provides continuity with the association of concept learning and an increasing, reflexive abstraction of students' conceptions.

The going conceptualizations of mathematical conceptions and concept learning are influenced by and based on the (constructivist) epistemologies offered up initially by Kant and Jean Piaget. The Kantian dogma of the a priori remained the dominating way of thinking mathematics into the last century; and Piaget adhered to Kantianism (Otte, 1998), though, because of his roots in biology, he abandoned the existence of a priori concepts such as those of time and space. But for Piaget, the individual subject still "abstracts" knowledge *away from* experience so that it achieves its experience-independent state (e.g., to engage in formal reasoning). Ideas therefore are no longer connected to the material practices from which they arise, such as when children learn fundamental mathematical operations using hands-on materials only to attain a formal world of mathematical operations. Both Kant and Piaget therefore are fundamentally Platonic thinkers in the sense that they conceive of mathematics generally and of geometry particularly as paradigmatic examples of knowledge that is independent of (abstracted away from) sensual experience, though always given in the form of representations that can be related to the things that we come to know through sensory experiences.

Immanuel Kant and the Construction of Mathematical Knowledge

Kant's account of mathematics is based on the activity of constructing mathematical objects in pure intuition (time and space). In yielding objects for mathematics, our intuition contributes in an essential way to the formulation of mathematical truths. To construct a concept means to represent the corresponding ideas a priori. The construction of a concept requires a non-empirical intuition, which consequently, as intuition, is a singular object, but nevertheless a construction of the concept: the universality of all sorts of ideas that belong to the same concept are expressible in some representation (Kant, 1968a). Thus, the I (self) is said to "construct" a triangle in representing the to-the-object-corresponding concept either through pure imagination in pure intuition or following it also on paper in empirical intuition. In either case, the representation occurs completely a priori, without borrowing the pattern from any bodily and embodied experience whatsoever. The individual figure is empirical and serves to express the concept in its generality. This is so because in empirical intuition, one always considers the action of constructing the concept—which is independent of many determinations, for example, the size, the sides and angles—and thereby abstracts from the differences between the things (e.g., *actual* triangles) that do not change the corresponding concept (e.g., "triangle").

In Kant's approach, the uppermost law of nature lies within ourselves, that is, within our reason, and that we cannot seek the general laws in nature, mediated by experience, but conversely, have to seek the general orderly nature of nature in the possibility of the senses and those in the mind-residing conditions of the possibility of experiences (Kant, 1968b). Kant realized that however abstract concepts might be associated—and therefore, however much they may have been abstracted from experience—with imagistic ideas, the purpose is to make them suitable for practical use even in the case that they have not been derived from such experience. For how could we otherwise associate sense and reference to conceptions if these were not grounded in some intuition—which always has to be somehow exemplified in possibility of an experience?

Kant distinguishes between transcendental and empirical concepts: The former are related to things (Ger. "Dinge") in general including those created by mind itself ("noumena"), whereas the latter can be related to phenomena ("phaenomena"), that is, objects of possible experiences. To exemplify his distinction in the context of geometry, for example, he uses the example of the cone: the "conic figure can be visualized without any empirical help, merely based on the concept, but the color of the cone requires that it had been given in one or another experience" (Kant, 1968a, p. 614). He also uses the example of a triangle that can be constructed on the basis of an a priori, non-empirical intuition by introducing a particular set of limits or boundaries into space in such a manner that a three-sided figure in a single dimension results. But in contrast to philosophers, mathematicians draw on representations, which allow them to view the concrete but only in a representation that they have constructed a priori. Again, Kant uses the concept of *triangle* as an example, pointing out that facing the question about the sum of the interior angles the philosopher ponders the term but does not achieve new knowledge whereas the mathematician constructs a triangle and exhibits for every one to see the classical proof that the sum is 180 degrees. An empirical proof or even the necessity for the angle sum to be 180 degrees would, so Kant, never be possible as it always leaves open the possibility that one finds another triangle in which this is not the case.

For Kant, concepts are organized hierarchically into genus and species, which indicate relative orders (Kant, 1968c). For example, in the context of my own research, where second-grade students are to learn concepts of three-dimensional geometry among others by classifying different objects (see chapter 4), there are rectangular solids, solids with curvature, pyramids, and so on. These solids can be grouped. Kant's theory implies that the highest concept is the one that cannot be a species (e.g., "solid"), and the lowest concept as the one that cannot be genus (e.g., "cube"). The extent of a concept is larger if it encompasses more things and therefore allows thinking more with it. The highest concept ("conceptum summum") is that from which nothing can be abstracted further without making it altogether disappear. A lower concept is not contained in a higher one, for it does contain within itself more than the higher one. But the former is contained underneath the latter, because the higher contains the epistemic ground of the lower. More so, a concept is not more encompassing than another because it contains more below itself—for this we cannot know—but insofar as it contains not only the other concept but additional concepts as well. What is valid to or what contradicts higher-order concepts is also valid for lower concepts that are contained within the higher concept; and conversely, what is valid for or contradicts lower-order concepts also is valid for or contradicts higher-order concepts.

Jean Piaget, Radical Constructivism, and Mathematical Knowing

Piaget and the (radical) constructivists took up Kant's approach in the *Critique of Pure Reason*. In the constructivist approach, the subject is constructing itself; and the emerging subject's (cognitive) structure is at the heart of the human apprehension of the world. Piaget reproduces Kantianism by emphasizing that human beings are actively involved in the world and have the potential for unlimited self-development. Radical constructivists went a step further emphasizing that cognizing beings do not have access to the world and its phenomena, that they are closed with respect to information, and that all that Being can do is testing concepts for the level of fit they provide. From the perspective of radical constructivism, knowledge, no matter how it is defined, "is in the heads of persons, and ... the thinking subject has no alternative but to construct what he or she knows on the basis of his or her own experience. What we make of experience constitutes the only world we consciously live in" (von Glasersfeld, 1995, p. 1). Even those conceptions that Kant thought to exist a priori, such as space and time, are, according to radical constructivists (in this following Piaget [1970]), the result of mental constructions that result from the individual's engagement with and as an extension of its lifeworld. That is, against Kant, Piaget argues that intuition has no place in mathematics generally and in geometry more specifically and that mathematical existence consists in the possibility of the defined objects, that is, in non-contradiction.

For Piaget, *abstraction* is a key element in the construction of conceptions and even more so in the construction of mathematical operations, as the mind engages in lifting structural properties from concrete experiences—for example, moving about and differently arranging pebbles and counting each configuration—to get to the properties in themselves, such as commutativity (in whichever order a collection of objects is counted, the total number remains constant). Piaget refers to this process as *reflective abstraction*. The concept denotes the fact that the mind/individual reflects on its/one's own operations with concrete objects and then extracts from these reflections general principles. What is taken away— that is, abstracted—yields a new object of a different order, an ideal object given to a form of reason of a different order, formal (abstract) reason.

With respect to the development of geometrical notions, for example, Piaget suggests that children—in contrast to the history of geometry, which moved from Euclidean metric geometry to projective geometry and to topology—begin by developing topological intuitions first. Among the early intuitions are those of ordering and dividing space. The latter is precisely the beginning for Kant, who supposes that given space as a priori, geometrical objects can be derived *without* any phenomenal experience through divisions of this space. The construction of an empirical (sensual) concept requires perceiving that which is unique and unchanging about a set of figures so that those can be left behind as the non-unique (that which they share) in the comparison.

Piaget shows that children of about four years of age, asked to copy a circle, a square, and a triangle, will draw shapes that all look about the same, that is, do not preserve the geometric properties of the different shapes. Asked to draw a cross, they do draw something that is or at least resembles a cross. That is, the children exhibit a topological intuition of closed curves that have been copied as closed curves versus open shapes that have been copied as open shapes. The children of this age also correctly copy relations, for example, small circle within a larger circle, small circle outside of the larger circle, and small circle lying across the circumference of the larger circle (Piaget, 1970). The difference between Piaget and Kant, therefore, is the presupposition of the latter that space is given a priori as the condition of experience as such, whereas the former shows that space itself is the result of experiences in the world. How the content of this world derives from appearances, however, is not solved in this approach (Henry, 2000).

Piaget and those following him differentiate between concrete concepts (constructed during the sensorimotor, pre-operational, and operational stages) and abstract knowledge (formal operational stage). For Kant, however, every concept is an abstract concept. We are held not to give privilege over the abstract or concrete use of concepts, for we recognize very little in many things with very abstract concepts and with concrete concepts recognize a lot in rather few things. Kant (1986c) suggests that all mathematical and experience-based concepts have to be constructed synthetically, so that—in contrast to some recent proposals—empirical concepts cannot be defined, because these are inherently open and unfinished. Only those concepts can be completely defined synthetically that are arbitrary, such as truly transcendental mathematical concepts.

Two Major Problems with and in Received Epistemologies

The approaches taken by Kant and Piaget after him lead to considerable theoretical problems. One of the most important one was the disjunction between concept(ion) and their realization in experience—a problem cognitive scientists refer to as the "symbol grounding problem." For mathematics educators, the same problem is articulated in the distinction between "knowing a concept" and "applying a concept." This problem disappears only when conception and embodied bodily experiences are theorized such that the order in the senses—and therefore the sense of concepts—arises from the senses themselves (Waldenfels, 1999). A second, related problem is that of the process of abstraction itself: To abstract a common property from some experiences, a property that characterizes the conception, the conception already needs to exist to be able to classify according to its characteristic property. Accordingly, the child develops conceptions grounded in invariant (logical) structures typical of adult rationality. It is evident that such a Piagetian approach cannot show how conceptions arise from "untamed" child-like experiences. It therefore also cannot show how the abstracting and abstract sciences could have emerged on cultural-historical grounds from primitive forms of thought.

CULTURAL-HISTORICAL ACTIVITY THEORY

Cultural-historical activity theory evolved in the Soviet Union beginning with the work of Lev Vygotsky and his students and followers. The theoretical foundation of the work lies in Karl Marx's philosophical work of dialectical materialism. Marx's fundamental critique of philosophy generally and of philosophies of the mind more specifically was aimed at their idealist nature. A prime example of idealism exists for him in the work of Immanuel Kant, who attempts to establish a science, whereby knowing and reasoning occurs independently of the world that we inhabit.[1] Marx considers the dialectic approach developed by Georg W. F. Hegel as giving an advantage over Kant, because it articulates how consciousness develops through repeated cycles of objectifying itself into things and signs, which it then takes up again and thereby subjectifies, individualizes.[2] Another conscious subject inherently recognizes these things and signs, because it sees in these the results of actions it could have produced itself. (Nowadays, neuroscientists tell us that this is so because of mirror neurons that fire both when we act and when we see someone else effectuating the action.)

Although Hegel constitutes an advance over Kant, Marx nevertheless accuses Hegel of being but the latest idealist in the history of philosophy. Marx therefore develops a materialist dialectical approach, in which consciousness does not merely develop in the manner Hegel indicates, but it develops by realizing the fullness of life in and through *productive material activity*. Such activity always has collective motives, such as farming to produce food or making tools to make farming possible. Consciousness (the ideal) and the material world—which includes the sounds we hear as words, writing, and other communicative forms—are two irreducible sides of the same coin. In any form of practical activity (even Gödel's proof or string theory requires concrete actions that others can observe [Livingston, 1986]), the ideal and material cannot be reduced to one another: they presuppose one another. On philosophical grounds, they have to have emerged together because otherwise there always will be contradictions between the relationship of thought and world (Henry, 2000). Most importantly, therefore, consciousness always is articulated in concrete activity and therefore in societal relations. It is precisely here, in and as societal relation, that Lev Vygotsky (1978), whose work is (though often denied in the West) through and through marked by dialectical materialism, finds the root of all higher cognitive functions. It is precisely here, then, that material practical activity becomes the leading edge of development: Children participate in activity prior to being conscious of the required knowledge, and it is in and through participation that they become conscious of this knowledge.[3] That is, with practical knowledge arises the possibility for understanding this knowledge theoretically. Knowledge does not come out of thin air or through children's singular abstraction from things and actions, as in Piaget's work, but rather consciousness evolves in and following participation in societal-material relations during concrete practical activity oriented toward the collective maintenance of life conditions.

Vygotsky's students and followers further developed the dialectical materialist approach to a fully-fledged theory of knowing and learning. Foremost among them was Alexei N. Leont'ev (1978), who differentiated between three different and mutually constitutive levels: activity, action, and operation. Practical, collective-life-sustaining activities, such as farming, manufacturing tools, or schooling have collective objects/ motives: they contribute to the reproduction and transformation of the collective (society). Individual and collective subjects (groups) realize activities through goal-directed actions. That is, the activity exists only because of the actions that realize it, but the actions are effectuated in view of the activity they are to realize: the two levels mutually presuppose and constitute each other. The *sense* of an action depends on the activity it realizes so that a particular action, in a different activity, has a different sense (Figure 1.1). The action of figuring out a best buy in a supermarket as part of the weekly shopping trip has a very different sense than the action of figuring out a best buy on a word problem in school. That is, what students do in school *are not* activities; these are better thought of as tasks that realize the activity of schooling, which itself has the motive of reproducing and transforming collective knowledge and practices of (immortal) society. The distinction between these two types of events is clear in the German and Russian languages where the theory originated, languages that clearly distinguish two concepts: societally motivated Tätigkeit/deyatel'nost' from mere being busy without a guiding motive, Aktivität/aktivnost'. Both concepts are translated into the same English word *activity*.[4] Both actions and activities have ideal and material dimensions that cannot be reduced to one another: The structures of the dispositions that underlie perception/thought and the structures of the material world stand in a mutually presupposing and constituting relation.

Conscious actions are not the lowest level of events that we observe: sequences of unconscious operations constitute an action. A farmer may decide to plow a field, but driving the tractor, shifting gears, keeping it so that furrow lies next to furrow does not require conscious deliberation from the experienced farmer. Each operation is produced just in time and as a function of the present material context. Similarly, speakers do not ponder each word or even grammar prior to speaking, but words come into and out of their mouths and as a function of what they have said so far. That is, the current state of the unfolding action provides a *referent* and condition for the next operation (Figure 1.1). Again, the relation is mutually presupposing and mutually constitutive: unconscious operations con-

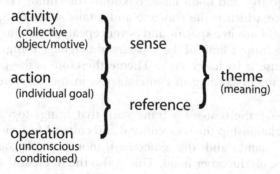

Figure 1.1.   The unit of "meaning" embodies dialectical relations between sense and reference, on the one hand, and between activity, action, and operation, on the other hand.

cretely realize an action, but the goal and the present state of the action constitute part of the conditions that determine the next operation.

Operations are events entirely bodily and embodied. Speaking requires the mouth and vocal cords, and shifting the gears in a tractor requires the movement of body, arm, and hand. Operations, however, are not singular: they are the result of mimetic reproduction (copying others) or have been conscious actions—that make sense only in societally relevant activity—that with familiarity and routinization sink below the threshold of consciousness and become unconscious operations. Thus, novice drivers of a tractor (vehicle) with a stick shift consciously attend to all aspects of shifting gears, when to push the clutch, when and how to disengage the gear by pushing the shift, and how to push the clutch again prior to engaging the next gear. Over time, what initially is an action with a goal (shifting gears) becomes an unconscious operation that does not even require tracking which gear one is in and whether and where to move the shift to reach the next gear up or down. Conversely, when there is trouble, an operation can be unpacked and its different constitutive moments can be brought into consciousness and therefore become actions again. Similarly, actions can become activities—such as when in early hunting societies tool making moves from the domain of each hunter into a self-sustained activity, where some individuals only make tools and make a living by exchanging the products of their labor against the meat that results from the hunting activity that others in the society realize.

Together, sense and reference constitute practical understanding in any concrete situation. Mathematics educators have come to use the adjective "meaningful" to denote when a human being has a practical understanding and a practical sense in and of some situation. The term, however, has been overused and scholars use it and the associated noun ("meaning") in very different ways leading to a total confusion just what sense the adjective and noun have. Bakhtine/Volochinov (1977) propose the term *theme*, which is the concrete and totally situated signification a word or sign obtains in a specific and never repeatable instant. The theme constitutes the upper limit of the capacity of language to signify, whereas dictionary sense is its lower limit. Theme therefore is the practical and participative understanding of consciousness in the concreteness of the given setting.

We now have the basis of a framework that brings together into an irreducible relationship the sociocultural and cultural-historical practices, on the one hand, and the embodied moments of cognition and consciousness, on the other hand. This is also the framework that I use in the epilogue to tease apart the critiques of Rafael Núñez's work, critiques that appear to be quite unjustified. As Figure 1.1 shows, actions are the lynchpin in the framework. Actions are observable and serve goals of the collectively motivated activity in progress. Their sense is tied to and arises from this role in concretely realizing the ongoing activity. The sense is something social, cultural, and historical through and through. This is why calculating a division by long hand, with a hand-cranked calculator, or with a slide rule to figure out a best buy in a supermarket makes little if any sense in a day and age with calculators and labels that give unit prices (if in small print). If a shopper were to engage in figuring out a best buy in this manner, others present might wonder and even consider the person as crazy (weird): "this makes no sense." But some 40 or 50 years ago, prior to the first handheld electronic calculators, the slide rule may have not raised an eyebrow whatsoever (other than others thinking that the user is an engineer or a physicist). At the same time, each action, even saying something and even just silently calculating a fraction in the mind, requires the body: all sorts of neurons and mirror neurons fire to produce operations that realize the actions concretely conditioned by the current state of affairs. Operations are bodily and embodied through and through. Concrete practical actions—even those of the mental or discursive type— therefore, are *irreducibly social and embodied at the same time*.

While reading the book, readers may keep this framework in mind, because it allows us to understand both the social and embodied moments of cognition even when the authors do not explicitly point to it. Thus, although Laurie Edwards or Rafael Núñez in their chapters by and large focus on embodiment aspects, what the students or the mathematician do and how they do it, once interrogated for sense, inherently reveals a cul- tural-historical dimension. Equally, whereas Paul Cobb and Carrie Tzou or Götz Krummheuer largely focus on the social dimensions of mathematical cognition, all actions and discourse students produce are inherently material and therefore embodied phenomena. The question to which the book aims to contribute in its entirety precisely is that of integrating these two perspectives so that we get a more holistic view of mathematical cognition as a simultaneously embodied and cultural- historical phenomenon.

NOTES

1. In a major work on the effort of Kant to separate knowing from the body, Jean-Luc Nancy (2008) calls the philosopher giant "Logodaedalus," the ultimate craftsperson of the logos, the pure mind.
2. Here I use the term consciousness for a very particular reason that allows me to link practical and material activity, on the one hand, and thought, on the other hand. A quarterback, for example, may be good at throwing the ball right to the wide receiver although he does not know the underlying

physics. But for scientists and mathematicians, what kind of knowledge counts is what is in consciousness.

3. In a moving account of his work with deaf and blind children, who heretofore only vegetated, the Russian psychologist Alexander Meshcheryakov (1974) allows these children to develop into normal adults by using Vygotskian/Marxian principles of development grounded in the idea that participation in societal relations leads to the development of higher-order cognitive functions.

4. This is a serious shortcoming equivalent to the case in physics of not distinguishing between heat (an energy), an extensive property of matter, and temperature, an intesnsive property of matter. In mathematics, an equivalent may be the conflation of the Boolean "and" (AND, ∍) and the algebraic "and" (+).

REFERENCES

Bakhtine, M./Volochinov, V. N. (1977). *Le marxisme et la philosophie du langage* [Marxism and the philosophy of language]. Paris: Éditions du Minuit.

Bereiter, C. (1985). Toward a solution of the learning paradox. *Review of Educational Research, 55,* 201–226.

Cobb, P. (1985). Two children's anticipations, beliefs, and motivations. *Educational Studies in Mathematics, 16,* 111–126.

Henry, M. (2000). *Incarnation: Vers une philosophie de la chair* [Incarnation: Toward a philosophy of the flesh]. Paris: Seuil.

Husserl, E. (1939). Die Frage nach dem Ursprung der Geometrie als intentionalhistorisches Problem. *Revue internationale de philosophie, 1,* 203–225.

Kant, I. (1968a). *Werke IV: Kritik der reinen Vernunft* (Vol. 2) [Works IV: Critique of pure reason, Vol. 2]. Frankfurt: Suhrkamp-Verlag.

Kant, I. (1968b). *Werke V: Schriften zur Metaphysik und Logik* (Vol. 1) [Works V: Writings on metaphysics and logic, Vol. 1]. Frankfurt: Suhrkamp-Verlag.

Kant, I. (1968c). *Werke VI: Schriften zur Metaphysik und Logik* (Vol. 2) [Works VI: Writings on metaphysics and logic, Vol. 2]. Frankfurt: Suhrkamp-Verlag.

Lakoff, G., & Núñez, R. (2000). *Where mathematics comes from: How the embodied mind brings mathematics into being.* New York: Basic Books.

Leont'ev, A. N. (1978). *Activity, consciousness and personality.* Englewood Cliffs, NJ: Prentice Hall.

Livingston, E. (1986). *The ethnomethodological foundations of mathematics.* London: Routledge and Kegan Paul.

Meshcheryakov, A. (1974). *Awakening to life: On the education of deaf-blind children in the Soviet Union.* Moscow: Progress.

Nancy, J.-L. (2008). *The discourse of the syncope: Logodaedalus.* Stanford, CA: Stanford University Press.

Otte, M. (1998). Limits of constructivism: Kant, Piaget and Peirce. *Science & Education, 7,* 425–450.

Piaget, J. (1970). *Genetic epistemology.* New York: W. W. Norton.

Von Glasersfeld, E. (1995) *Radical constructivism: A way of knowing and learning.* London: Falmer.

Vygotsky, L. S. (1978). *Mind in society: The development of higher psychological processes*. Cambridge, MA: Harvard University Press.

Waldenfels, B. (1999). *Sinnesschwellen: Studien zur Phänomenologie des Fremden 3* [Sense boundaries: Studies in the phenomenology of the foreign/strange 3]. Frankfurt: Suhrkamp-Verlag.

PART A

MOVING AND TRANSFORMING BODIES IN/AS MATHEMATICAL PRACTIVE

EDITOR'S SECTION INTRODUCTION

Wolff-Michael Roth

Embodiment, cognition, and the relative motion of bodies—with respect to other bodies particularly and the material world generally—are at the very heart of human consciousness, they *are the heart* of consciousness. Without the proximity of the human body to other human bodies and without the movements of the body with respect to the world, we would not be able to experience ourselves as different from other selves, different from the world, different from the signs (representations) that we use to denote other parts of the world (Levinas, 1998). The movement of the human body and human flesh is the *condition* for anything like cognition, memory, apprehension and comprehension of the Self, the relation between thought and the world, and so on (Henry, 2000). The four chapters in this section focus on the movement of bodies and body parts and the translations they undergo in the conceptualization and reconceptualization of geometrical (graphical) objects and understanding.

One of the questions seldom asked in the literature on mathematical learning is how the two perspectives, one that focuses on the bodily, embodied nature of mathematical cognition and the other that focuses on its social nature, can be thought together. Cultural-historical activity theory, with its simultaneous orientation toward embodied operations and societal (social) already integrates the two perspectives, allowing us to theorize the intricate relationships not only between individual and social cognition but also, not emphasized in this book, other aspects of

Mathematical Representation at the Interface of Body and Culture, pp. 19–25
Copyright © 2009 by Information Age Publishing
All rights of reproduction in any form reserved.

mathematical activity such as motivation and emotions (e.g., Roth, 2007). Another avenue to theorizing the material body and the social aspects of knowledge together has emerged in a line of work that began with phenomenological sociology (e.g., Alfred Schutz). This line, on the one hand, has developed into ethnomethodology and conversation analysis, and, on the other hand, has found its expression in early forms of dialectical theories in which societal fields and individual dispositions are complementary aspects of the same situations. The first line of work has been realized, for example, in the ethnomethodological studies of mathematicians (Livingston, 1986) whereas the second line of work centrally focused on mathematical cognition in the everyday world (e.g., Lave, 1988). In both lines of work, great attention is paid to the various ways in which mathematical cognition is exhibited, which involves not only mind, but also and especially so the body; and because mathematical cognition is exhibited, it inherently is *for others*, and therefore, a collective phenomenon even though singularly expressed in and by the performance of individual bodies.

In cultural-historical activity theory, there is a close linkage between the embodied moments of practical work, on the one hand, and the cultural-historical moments of practical work, on the other hand. In the chapters that make this first part of the book, the activity level and how it mediates is not explicitly addressed. Thus, the children (chapter 4), students (chapter 2, 3), and physicists (chapter 5) are in particular settings where they participate in making school or making a think-aloud protocol for research purposes. These participants are not merely oriented to one another but are oriented more generally to the collective activity in which their actions make sense. Thus, for example, the physicists in chapter 5 do what they do in a way so that the cognitive scientists asking them to participate in the think-aloud protocol obtain data suitable for analysis. Whether their actions make any sense in their own laboratory work remains an empirical question; importantly, these physicists collude with the cognitive scientists in making the think-aloud possible but their talk and therefore the analyses do not explicitly raise this issue. Similarly, the children and students in the other chapters collude to make school happen in the way it does, so that their actions make sense in the context of schooling whereas they might not make any sense whatsoever in their mundane world outside schools. (There, the same actions might be taken as signs of nerdiness.)

In all four chapters of Part A, participants have to perceive physical entities (geometrical objects, graphs). Whereas often unattended to as a special problem in mathematics, perception itself is a phenomenon that straddles the body and society. This can therefore be captured in cultural-historical activity theory, where an explicit relation is made between the unconscious bodily operations, on the one hand, and the goal-oriented actions, on the other hand. Unconscious operations, for example, include the perceptual processes that bring some aspect into awareness. Thus, in chapter 4, Jean-François Maheux and colleagues begin with observations concerning the perceptions of a drawing that—in the present cultural-historical context—is seen as depicting a cube. For the purpose of this discussion concerning the embodied and cultural moments of perception, I present this cube together with two others (Figure A.1).

Upon first sight, the first drawing (Figure A.1a) might be seen as a cube, if it is seen as a cube at all, viewed from slightly above and from the front right. But it might also be seen as a cube from below and front left. Yet there is nothing inherent in the drawing that makes us see the drawing as a cube rather than a bunch of lines. That we see it as a cube (if we see it as such at all) is both a cultural and an individual-developmental phenomenon. Very small children will not see a cube; and whether the members of Amazonian tribes that have not had contact with Western civilization see it as a cube would have to be tested empirically. To find out what is at the origin of the experience of seeing the cube, we can conduct phenomenological explorations. These show what the underlying unconscious operations are that make us see the drawing as a cube. We may do so, for example, by exploring how to switch quickly back and forth from the cube seen slightly from above to the other one seen from below. The full exploration would take more space than I have here. Let it be said that if one moves the eye to the lower left inner corner that appears within the set of lines and then moves toward a non-present vanishing point to the left ("along the surface"), the cube becomes instantly apparent. Similarly, focusing on the equivalent corner further up and to the right and then moving along the line "backward" to a non-existing vanishing point allows us to see a cube seen from the top. That is, unbeknownst to our consciousness, the movement of the eye

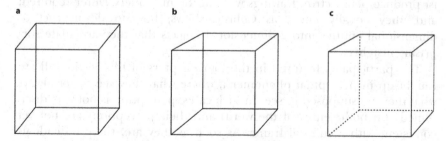

Figure A.1.   Two-dimensional drawings that can be seen as cube. a. Necker cube. b. Cube drawn with vanishing point. c. A cube with "hidden edges" drawn as dotted lines.

toward a non-existing vanishing point creates one or the other experience. If the eye does not make this movement, then the cubes do not appear and the lines remain lines on a flat surface. Experiences with cubes as cubes and drawings of cubes in a social context allow children to see the drawings as 3-dimensional cubes. In fact, newly born children do not see perspective at all so that the world appears flat to them. They perceive perspective after beginning to move about (or to move parts of the bodies, including the eyes) and experience the changes in the visual field coordinated with their own movements.

The cubical nature is more determined in the other two instances (Figure A.1b, c), which are drawn to have a single vanishing point (take a ruler and follow the lines backward, they will intersect in one point). If the same exercise as before is done with the center cube (Figure A.1b) but for the inner corner in the lower center left, then the cubical nature disappears and we see a part of a pyramid narrowing toward the front. The cube no longer is a cube when the vanishing point is changed. When the lines that in a solid cube would be hidden are drawn in dotted lines (Figure A.1c), then the eye is almost forced into a viewpoint from above.

That perspective drawings are cultural and historically contingent can be seen from the changes that drawings have undergone from the early cave drawings and paintings, via the flat paintings where, similar to children's drawings, all body parts are seen flat as in Egyptian painting and drawing (Figure A.2), to the fully perspective drawings in the Renaissance. Drawing and seeing perspective involves work, as a well-known Albrecht Dürer drawing illustrates, which shows the artist viewing a naked woman via a miniature obelisk (functioning as a fixed point) and a grid posed between the obelisk and the naked woman. That is, to see in perspective he required an apparatus that mapped the three dimensions onto two-dimensions but in a way so that the eye can recover the three dimensions. This, in fact, requires a transformation that no longer maps the source space (3D) in a one-to-one manner onto the target space (2D). Today, artist produce perspective drawings without the machinery Dürer required; and, they equally know, as Cubism shows, how to deconstruct 3-dimensional entities into 2-dimensional objects that they articulate surprising insights.

The participants featured in the four chapters of this section all are caught up in perceptual phenomena, where what they see is not always what they are supposed to see. In such cases, perception is not yet "disciplined" (in both senses of the word) and their perceptions are not yet consistent with a cultural frame. As soon as they are, these individuals perceive entities in the same way that other experienced practitioners in the field do. Thus, the children in chapter 4 are in the process of learning that cubes are part of the same class independent of color and size; that

Figure A.2.   During the times of the Egyptian pharaohs, all parts of the body were drawn frontally, so that the bodies look twisted.

is, they learn to see cubes consistent with a mathematical (geometrical) classification rather than consistent with other experiences and possible ways of classifying objects. It is not surprising that they initially tend to classify, for example, cubical objects by size, because on the playground, they likely do very different things with cubical objects of 1-m by 1-m by 1-m dimensions than if the same dimensions are 10 cm. Thus far, therefore, I show how perception is an embodied phenomenon with virtually no connection to culture.

But in the cultural-historical activity theoretic framework, people always are caught up in some activity as part of which they communicate. In communicating, we explicitly address others for whom we speak. Thus, and most explicitly in my own chapter 5, the collection of chapters in this first section work out a transition from perception as an individual embodied phenomenon to perception as a cultural phenomenon. All of these chapters, therefore, though beginning and focusing on embodied aspects really find themselves at the juncture with collective effects, as they are all involved in communication. A full articulation of the social and cultural moments of knowing is presented in the chapters of the second section, where we find less attention to the embodied moments of mathematical cognition.

In chapter 2 ("Transformation Geometry from an Embodied Perspective"), Laurie Edwards uses a theory of embodied cognition to interpret results from a series of studies of students interacting with a computer microworld for transformation geometry. She shows how participants from three different groups (middle school, high school and undergraduate students) demonstrate very similar naïve interpretations and "misinterpretations" of Euclidean transformations when they use the microworld. The theory of embodied cognition and the related field of cognitive linguistics that she draws on allow her to make sense of these results and to contrast the students' understandings with how geometry is construed in contemporary mathematics.

From a cultural-historical perspective, one of the crucial developmental steps in the mathematical study of motion was the creation of a systematic system of reference. The creation of such a system required, in particular, a mathematical reconceptualization of space and time. The result of this endeavor was what is now called a Cartesian plane. However, graphs—that is to say, the signs of a Cartesian plane—rest on a sophisticated manner of signifying that, from an ontogenetic viewpoint, is far from transparent to novice students. In chapter 3 entitled "Moving Bodies: Linking Experiential and Mathematical Spaces," Luis Radford presents tenth-grade students in their attempt to make sense of a graph representing the relative motion of two bodies. In particular, Radford discusses, from the point of view of cultural semiotics, the students' gestural, kinesthetic, symbolic, and discursive activity as they disentangled the various experiential (phenomenological, imagined, kinesthetic) spaces and linked them to the mathematical conceptual one.

In the early years, learning mathematics involves exploring both the material world and corresponding ideas with our bodies in a way that one can culturally associate with the field of mathematics. In chapter 4 entitled "What Makes a Cube a Cube? Contingency in Abstract, Concrete, Cultural and Bodily Mathematical Knowings" Jean-François Maheux, Jennifer Thom, and Wolff-Michael Roth discuss how the physical body and cultural practices are constitutive parts of each other, and illustrate how this plays out in the learning of mathematics. More precisely, the authors draw on specific case exemplars in a second-grade geometry classroom to focus on mathematical abstraction. Maheux et al. articulate that abstractions are sociocultural creations and possibilities concretely realized by individuals, in and through their sensorimotor activities (bodily knowing), to participate in exploring the world and ideas (cultural knowing). They demonstrate that in the transactions with the world, involving students' senses, students make sense of their experiences realize an ongoing process of articulation of mathematical abstraction that increases their room to maneuver in the back and forth movement between culturally mediated and unmediated actions. At the interface of the body and culture, this shows how mediational artifacts create possibilities to develop ways of mathematically being in the world, connecting learning at the embodied level with contributing in a mathematics classroom practice, and beyond that, participating in a cultural form of knowing.

In the cognitive science and mathematics education literature on graphs and graph-related skills (interpretation, production), it is assumed that certain features (e.g., slope, height) inherently stand out, which implies that those asked to interpret a graph are assumed to be addressing the sense of these features. The epitome of this approach can be seen from cognitive science research that proposes mental models that include iconic images of the graphs in the human brain (neurons). On the other hand, there are a number of studies among scientists that show how the visibility of graphical features itself is the outcome of a collective process, which elaborates what it is that can be seen objectively. In "Embodied Mathematical Communication and the Visibility of Graphical Features" (chapter 5), I (Wolff-Michael Roth) investigate, using a method grounded in conversation analysis and ethnomethodology, the process of how graphical features become collectively visible and objective by analyzing the work of pairs of scientists asked to interpret infield and out-of-field graphs. Pairs have been chosen rather than individuals in think-aloud sessions because, in attempting to complete the requested task, the participants make available *for* one another what is required to remain aligned in solving the problem at hand. The study shows that perception includes an essentially passive (rather than intentional) moment, where features have to be understood as given to the individual perceiver. Any initially individual way of seeing a feature becomes objectively available in and for the community through the embodied communication (involving [deictic, iconic] gestures, body orientations, and prosody). For both features of the graph interpretation sessions, the products are better understood as emergent rather than as the outcomes of intentional processes.

REFERENCES

Henry, M. (2000). *Incarnation: Vers une philosophie de la chair* [Incarnation: Toward a philosophy of the flesh]. Paris: Seuil.

Lave, J. (1988). *Cognition in practice: Mind, mathematics and culture in everyday life*. Cambridge, England: Cambridge University Press.

Levinas, E. (1998). *Otherwise than being or beyond essence*. Pittsburgh, PA: Duquesne University Press

Livingston, E. (1986). *The ethnomethodological foundations of mathematics*. New York: Routledge.

Roth, W.-M. (2007). Emotion at work: A contribution to third-generation cultural-historical activity theory. *Mind, Culture, and Activity, 14*, 40–63.

CHAPTER 2

TRANSFORMATION GEOMETRY FROM AN EMBODIED PERSPECTIVE

Laurie D. Edwards

The theory of embodied cognition is utilized to interpret results from a series of studies of students interacting with a computer microworld for transformation geometry. Participants from three different groups (middle school, high school and undergraduate students) demonstrated very similar naive interpretations and "misinterpretations" of Euclidean transformations when they used the microworld. The theory of embodied cognition and the related field of cognitive linguistics are used to interpret these results and to contrast the students' understandings with how geometry is construed in contemporary mathematics.

INTRODUCTION

During its existence as a formal discipline, the field of mathematics education has drawn from different theoretical schools and core disciplines in order to carry out its investigations of mathematical thinking and learning, and has also generated new theories and explanatory paradigms of

Mathematical Representation at the Interface of Body and Culture, pp. 27–44
Copyright © 2009 by Information Age Publishing

its own. There have been analyses of children's mathematical activity from the perspective of information processing psychology and Piagetian constructivism. Mathematicians and philosophers have examined the nature of mathematics itself, whether considered as a transcendental domain or a human creation. During the past decade, a "turn to the social" within mathematics education has taken place, as researchers seek to understand how individual knowledge is related to the fundamentally social nature of human beings, and how mathematics itself can be viewed as a specialized form of social discourse. In general, one can perceive an overall broadening of boundaries within the field, from an early focus on individual cognition and problem solving to growing perspective that sees cognition as a phenomenon that is situated in specific contexts and distributed across people and artifacts.

A recent "turn to the body" within cognitive science has further expanded the investigation of the nature of cognition, including mathematical thinking. Research attending explicitly to the role of the body in conceptualizing, doing and learning mathematics is beginning to appear, in conjunction with a reframing of knowledge as something that is dependent on the mind, rather than an objective feature of the universe. An embodied view of mathematics is not incompatible with a social one; in fact, both share a non-objectivist view of mathematics and knowledge, as evidenced by Anna Sfard's (2000) statement that "Knowledge viewed as an aspect of discursive activity is no longer a disembodied, impersonal set of propositions, the exact nature of which is a matter of 'the true shape' of the real world; rather it is a human construction" (p. 161). Embodied cognition, however, attempts to describe how the "shape of the world" and our activities within it provide a grounding for discourse and help to make that discourse mutually intelligible among its participants.

The theory of embodied cognition rejects a model of cognition as an abstract process of symbol manipulation, equally suitable for implementation on a computer as in the human brain. Instead, the embodied cognition perspective holds that,

> Thinking and learning are ... situated within biological and experiential contexts, contexts which have shaped, in a non-arbitrary way, our characteristic ways of making sense of the world. These characteristic ways of understanding, talking about, and acting in the world are shared by humans by virtue of being interacting members of the same species, co-existing within a given physical medium. (Núñez, Edwards, & Matos, 1999, p. 47)

Research in embodied cognition examines how both evolutionary constraints and individual bodily experience provide a foundation for the distinctive ways that humans think, act, and speak.

An approach to understanding mathematical ideas from an embodied perspective is compatible with recent investigations of the neural, developmental and biological bases of quantitative thinking, in both humans and other animals. It is also closely related to the field of cognitive linguistics, which seeks to identify fundamental conceptual mechanisms, such as unconscious metaphor, image schemata, and conceptual blends, that underpin and provide structure to a wide range of human thought (e.g., Fauconnier & Turner, 2002). Cognitive linguistics does not view language as a formal abstract domain, but rather as a system whose structure is inextricably rooted in human experience with such basic physical distinctions as up and down, inside and outside, paths and goals. The theory of embodied mathematics applies principles and findings of cognitive linguistics and embodied cognition to understanding mathematics as a human construction. An embodied approach to understanding mathematics seeks to frame mathematical thinking and learning not as a "special case," with characteristics that set it apart from other kinds of thinking and which require special mechanisms, but as a phenomenon that shares properties and explanatory principles with other kinds of human (and even animal) cognition.

In this chapter I describe the theory of embodied mathematics and utilize an embodied perspective on mathematical thinking to analyze data from students' experiences in learning transformation geometry. The thesis of the chapter is that students' understanding of Euclidean transformations can be understood fruitfully in terms of embodiment in general, and by using the analytic tools of conceptual linguistics in particular. The results of this analysis demonstrate that students' conceptualizations differ in interesting and non-arbitrary ways from those reflected in the discipline of formal mathematics. The students' thinking is not simply an incomplete or erroneous version of expert knowledge, but instead draws from a completely different set of conceptual sources.

THE THEORY OF EMBODIED MATHEMATICS

Embodied cognition states, in essence, that ideas emerge through the interaction of the mind with the world, and that each of these elements (the mind and the world) offers constraints on, and affordances for, the kind of ideas human beings are able to construct. Since human beings do not exist in isolation, these constructions take place within particular social contexts, and so cognition is also fundamentally social and situated. As applied to mathematics, the theory of embodied cognition, according to Lakoff and Núñez (2000), holds that

> Mathematics, as we know it or can know it, exists by virtue of the embodied mind. All mathematical content resides in embodied mathematical ideas. A large number of the most basic, as well as the most sophisticated, mathematical ideas are metaphorical in nature. (p. 364)

It is important to note that the theory of embodied mathematics is not an account of the properties of mathematics considered as a set of transcendent truths but is about the mathematics that humans are physically able to conceive of, communicate about, and practice. At the same time, the theory does not deny the objectivity of the physical world. In fact, the theory proposes that the nature of (humanly conceived) mathematics arises, directly or (in most cases) indirectly, from the way that the embodied mind exists in and interacts with the physical world.

As we interact with it, the world behaves in a fairly predictable way, and the foundations of mathematics, in a conceptual rather than logical sense, are based on the experience of manipulating and interacting with this predictable world. If a collection of two objects is put together with another collection of two objects, the result will be a collection of four objects, not one of five or three. Human infants behave as if they have some awareness of this kind of predictable behavior. Recent experiments have shown that babies will increase their attention to situations in which these "arithmetic" principles are apparently violated, although this awareness extends only to the very smallest of collections, namely, single objects combined with single objects (e.g., Dehaene, 1997). The ability to subitize, to keep track of small collections of objects, seems to be part of a set of innate or virtually innate abilities humans share with a few other kinds of animals, for example, chimps and parrots. These basic abilities serve as a starting point for the development of the rich and complex system that mathematics has become in various cultures, over the centuries. But this starting point is very restricted, and work in embodied mathematics is also concerned with identifying conceptual mechanisms and building blocks that allow humans to extend from these basic biological capabilities to construct the entire range of humanly conceived mathematical ideas.

COGNITIVE LINGUISTICS:
CONCEPTUAL MECHANISMS AND PRIMITIVES

The field of cognitive linguistics describes a set of basic mechanisms and structures that can account for the ways that people think and talk about a wide range of phenomena, from politics to human relationships to mathematics. These primitives and mechanisms include prototypes, image schemas, aspectual concepts, conceptual metaphor, conceptual blends, and metonymy. Within the theory of embodied mathematics, conceptual metaphor is an important (but not the only) mechanism proposed to account for how mathematical understandings are connected to the world and to each other. Conceptual metaphor is an unconscious mapping between a well-understood source domain to a less-well-understood target domain, one which carries over aspects of the inferential structure of the source domain into the target, thus permitting the construction of new ideas out of old. An example from outside of mathematics can be found in the way that we talk about scholars "building" a new area of research. The source domain in this case is physical construction, and the target domain is abstract knowledge. The fact that we are familiar with what it means to build something physically provides a way to think about the process of creating and sharing new knowledge. (This metaphor is found in other common ways of talking about intellectual products, as when we say, "His argument has no foundation" or "She extended her ideas into a new area.")

Within mathematics, a simple example of conceptual metaphor would be the "arithmetic as object collection" metaphor, where our common, embodied experience of grouping or collecting objects serves as the source domain for constructing the arithmetic of natural numbers. A given target domain, however, might have more than one source domain. For instance, arithmetic can also be conceptualized as constructing an object from different parts, or as measuring physical segments, or as motion along a path. Each of these metaphors would bring different entailments. For example, only in the "motion along a path" metaphor is zero an inherent part of the basic metaphor. A "collection" of size zero is not an inherent, natural part of the object collection source domain. Rather, the metaphor has to be extended to allow for this essential aspect of whole number arithmetic. It should be noted that although much of the richness and development of mathematics as a discipline may be related to determining the relationships among different metaphorical mappings, the fact that students and teachers may, without realizing it, be working from different source domains when discussing a mathematical topic can result in misunderstandings and ineffective instructional strategies, as I discuss below in the case of transformation geometry.

The conceptual metaphors for arithmetic are examples of a type called "grounding metaphors." Grounding metaphors draw from the physical world and direct bodily experience for their source domains; in this sense, it is easy to see how mathematics can be "embodied." However, embodiment does not imply that every mathematical idea derives directly from the physical world or physical actions (for example, the use of mathematical "manipulatives" by young students). Most mathematical concepts and processes, according to the theory, are constructed through mappings between and among existing mathematical ideas. Thus mathematics, as a human construction, can be thought of as a complex and sophisticated layering of metaphors and other kinds of mappings, ultimately, though not proximally, grounded in the physical world.

The types of metaphors used to build more sophisticated levels of mathematics by mapping between existing subdomains are called "linking metaphors." As an example, conceptualizing a fraction as a ratio of two whole numbers (rather than as a "piece of a whole") would be a linking metaphor, since each domain has its own structure, with its own entailments. A third type of metaphor, also not based on physical experience, is the redefinitional metaphor, defined as "metaphors that impose a technical understanding replacing ordinary concepts" (Lakoff & Núñez, 2000, p. 150). A great deal of the analytical work carried out by Lakoff and Núñez and others is concerned with isolating and describing plausible linking and redefinitional metaphors and conceptual blends that give rise to different mathematical concepts and subdomains. An example, important in the analysis of the work with transformation geometry, are the metaphors that redefine numbers as sets, or a space as a set of points; these metaphors are described in more detail below.

Conceptual metaphor is only one of the building blocks proposed in the theory of embodied mathematics. Conceptual blending, also called conceptual integration is another. Blends consist of mappings that draw from more than one source or input domain to allow the construction of a new domain, the "blend," which is isomorphic to none of the inputs, but which draws from the inferential structure of each. An example would be the "numbers as points on a line" blend, where drawing from previously-constructed understandings of both numbers and lines, new entities, "number-points" are created that have characteristics of both. Another class of very important cognitive primitives consists of image schemas, which are largely unconscious structurings of mathematical entities based on visual and spatial experiences of how the world works. The container schema is an example, in which our understanding of, and experience with, how physical containers work partially grounds our understanding of how classes or sets work. That is, our actions as very young children allow us to build the image schema associated with something being "in" or "inside of" something else (the container). These image schemata, in turn, support our later understanding of the inclusion of elements in a set (with the understanding of the abstraction, set, built on our knowledge of physical containers).

In the remainder of this chapter, the tools of embodied cognition and cognitive linguistics, as well as specific findings from the work of Lakoff and Núñez, are applied to understanding certain results from a series of investigations of students' learning with a computer microworld for transformation geometry.

TRANSFORMATION GEOMETRY AND THE TGEO MICROWORLD

Transformation geometry has formed a part of the mathematics curriculum in Britain for some time, having been introduced in textbooks in the United States somewhat later. The topic is sometimes introduced in U.S. textbooks as "motion geometry," with transformations described (and illustrated) as physical motions such as "flips," "turns," and "slides." A more formal definition of a transformation of the plane states that

> A one-to-one correspondence of the set of points in the plane to itself is a *transformation of the plane*. If point P corresponds to point P', then P' is called the *image* of P under the transformation. Point P is called the *preimage* of P'.
> (Long & DeTemple, 2003, p. 826)

In contrast to a more traditional approach to geometry, in which students investigate properties of figures such as lines, circles, polygons, and so forth, transformation geometry takes mappings, or transformations, of the plane as the objects of study. The study of transformation geometry in school typically addresses definitions and characteristics of individual transformations (for example, translations, rotations, reflections, and dilations), composition and inverses of transformations, and, in some curricula, matrix representations for transformations.

Figure 2.1 illustrates the set of transformations instantiated in TGEO, a computer microworld for transformation geometry that I developed some time ago (Edwards, 1997). The microworld includes three Euclidean transformations, or rigid motions, of the plane (named Slide, Rotate, and Reflect in the microworld), one dilation (Scale), and simple local versions of rotation and reflection (Pivot and Flip). TGEO is a multiple representation environment; in order to carry out a specific transformation, the user types its name, along with the required inputs, into one window of the microworld, and views the results of the transformation in a second graphics window. The graphics window shows the pre-image and image of each transformation or sequence of transformations, utilizing a block letter "L" as a concrete display of the effect that the mapping has on the whole plane.

The microworld, which was designed initially for use with middle school students, utilizes a simple set of commands with inputs that specify the particular mapping. The Slide transformation (translation) takes two

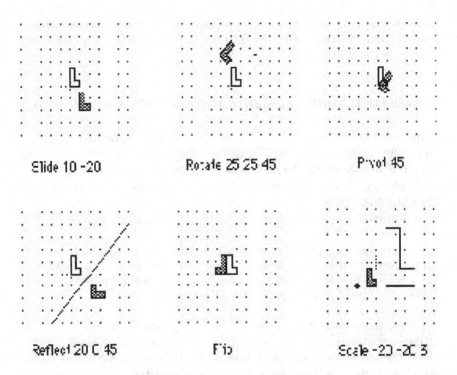

Figure 2.1.   The TGEO microworld.

inputs; the first gives the horizontal displacement and the second the vertical displacement of the mapping of the plane from itself to itself. Thus, "Slide -30 20" would move the plane (and the L-shape) 30 "steps" to the left of the origin, and 20 steps up. Rotate requires three inputs, with the first two used to specify a center point anywhere on the plane and the third the amount to rotate the plane around that fixed center. For example, the command "Rotate 20 20 90" would implement a clockwise quarter-turn of the plane around a fixed point at (20, 20) (the clockwise direction for rotations was chosen for consistency with the middle school students' previous experience with the programming language LOGO).

Reflect also takes three inputs, which are used to specify an arbitrary mirror line somewhere on the plane, over which the entire plane is mapped. Again, the conventions for the symbolic representation were chosen to build on the students' LOGO experience. The first two inputs specify a fixed point on the plane, and the third input gives the heading for a unique line through that point, with "0" representing a "north" heading (again, the default for the LOGO turtle). So the command "Reflect 0 20 90" would result in a reflection over a horizontal line which is 20 "steps" above the origin, since the mirror line for this reflection passes through the point (0, 20) at a heading of 90°.

The default behavior of the microworld is for each successive transformation to act on the state of the plane resulting from the previous transformation; thus, composition of operations consists simply of entering two or more transformations in succession. The microworld also included certain built-in activities, including a "matching" game in which the object was to utilize the transformations to superimpose an L-shape that started at the origin onto a second L-shape placed at a random location on the plane.

ETHNOGRAPHIC CONTEXT OF DATA

In this chapter I re-examine data collected during a set of studies with 11–15-year-old students using a computer microworld for exploring transformation geometry. Additional data were gathered in a study of adult undergraduates, prospective elementary school teachers. In the first study, three Euclidean transformations, translation, rotation and reflection, were introduced to a class of sixth-grade students. In developing the concepts of the individual transformations, sheets of overhead transparencies were used to illustrate the motions or mappings. Then 12 students (9 boys, 3 girls) worked in pairs for a period of five weeks on a series of problems and explorations using the computer microworld. For example, in one of the first activities, students played the matching game described above in which they utilized the transformations to superimpose one L-shape onto another. Later, the students determined inverses for each transformation, and investigated composition of transformations, inductively discovering the single transformations that are equivalent to pairs of translations, rotations, or reflections.

A similar study was carried out two years later with a group of 10 high school students (4 girls and 6 boys, from 15 to 16 years of age). In this study, the students were introduced to the transformations in pairs, rather than in a whole group, using drawings on a sheet of paper to illustrate the transformations. The students carried out the same sequence of activities as the younger students, with more attention given to investigating composition of functions. Finally, 14 college undergraduates participated in a study in which they first had to predict the outcomes of several transformations on paper, and then utilize a version of the microworld to check their answers. Although there were some methodological differences across these studies, overall, the results of all three investigations, whether they involved middle school, high school or adult students, were remarkably similar. None of the participants had studied transformation geometry in school, yet their non-standard interpretations of certain transformations were the same, regardless of age. These similarities are described in more detail below, but the consistency across both younger and older learners in how they interpreted the transformations raises the question of the source of their initial understandings of the domain— were the participants drawing from the same source domains in constructing their understanding of geometric transformations? In order to guide the development of effective instruction, it would be important to understand how these initial conceptualizations differ from the way that transformation geometry is defined within contemporary mathematics., and how to bridge these initial interpretations with the mathematically standard version of the transformations.

SELECTED RESULTS

Selected results from the studies involving the transformation geometry microworld are presented here, focusing on situations in which the students' work diverged from the formal mathematical conceptualizations of geometric transformations. As noted above, a formal mathematical definition holds that transformations are mappings of the plane to itself. In contrast, students often worked with the transformations as motions rather than as mappings. This conceptualization of transformations as motions was displayed in several different ways. When asked about how translations worked, one middle-school student stated "You tell something where you want it put and then it slides it over to that place" (Edwards, 1997, p. 195). This description not only emphasizes the motion of the shape on the screen, it also focuses on the final location of the shape, rather than on the relative change in coordinates, which is the aspect most salient when specifying the parameters of a translation, and when considering a translation as an object, rather than as a one-time action or process

All of the students developed a strategy for playing the matching game that was consistent with a "motion" view of the transformations. The goal of the game was to utilize as few transformations as possible in order to superimpose two L-shapes. If the screen had been viewed as a plane and the transformations as mappings of the whole plane to itself, executing one or at most two transformations could often solve the game. However, most of the students did not find these optimal solutions. Instead, they discovered a less efficient but reliable strategy labeled (by the researcher) as "Slide-Pivot-Flip." As the label suggests, this strategy consisted of sliding one of the shapes to the location of the second, then Pivoting in place, and finally (if necessary), Flipping in order to end up with the correct handedness. This strategy is indicative of viewing transformations not as mappings of the plane but as motions of objects on the screen. In general, the students preferred using the simpler, local versions of the transformations (Pivot and Flip) rather than the more global Rotate and Reflect; for the latter transformations, it is more difficult to see them as motions, because, unlike with Pivot and Flip, the center of rotation or the reflection line are often "at a distance" from the shape itself.

THE ROTATE "BUG"

A final result indicative of a "motion" view of the transformations was demonstrated in a naive interpretation of rotation on the part of a number of students, an interpretation that I called "the Rotate Bug." Rotate requires that the user enter three parameters. The first two are used to specify a center point at any arbitrary location on the plane, and the third to specify the angle through which the entire plane should be rotated, with the center point held invariant. A number of students expressed surprise when they first used the Rotate command. Rather than anticipating a mapping of the entire plane, the students expected that this command would Slide the shape to the given center point, and then Pivot it around it (see Figure 2.2). They seemed to have difficulty seeing rotation as occurring "at a distance" from the object, that is, in seeing the entire plane "carrying" the L-shape as it turned around a central point. This "misconception" cropped up among approximately a third of the students in the first study with middle-school students, and among a smaller number of high school and adult students.

In most cases, when the students were surprised by or had difficulty with understanding something that happened in the microworld, they were able to self-correct, simply by trying additional examples, and discussing it among themselves. However, with the Rotate Bug, the students did not find it easy to correct their own understanding by further exploration, and generally needed an explicit explanation from the researcher of how the Rotate command actually worked.

One of the conclusions of the initial study involving middle school students was that,

The students' understanding of transformation geometry as it developed in the ... game was still relatively unsophisticated. Their model of a transformation fell somewhere between, on the one hand, the notion of a direct, physical manipulation of an isolated shape and, on the other hand, the mathematician's concept of a specified mapping from the plane to itself. One of the most important missing aspects of the students' understanding

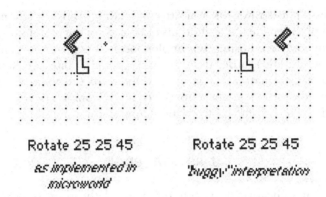

Rotate 25 25 45
*as implemented in
microworld*

Rotate 25 25 45
"buggy" interpretation

Figure 2.2.   The Rotate "bug."

of the transformations was in seeing them not as motions of an isolated object, but as a change involving the whole plane. (Edwards, 1989, p. 279)

In analyzing these results, I place them into a broader context that includes an examination of how mathematicians view geometry, and how the "official" conceptualization of this domain has changed within the broader development of the field of mathematics itself.

GEOMETRY AND THE DISCRETIZATION PROGRAM

Prior to the nineteenth century, geometry was conceptualized very differently than it is within advanced mathematics today. From the time of Euclid and before, geometry was seen as the investigation of the properties of objects in the plane and in 3-dimensional space (for example, points, lines, triangles, spheres, etc.), and the discovery, through both inductive and deductive means, of truths about these objects. During this period, space, including the plane, was conceptualized in an intuitive way, with the property of being what Lakoff and Núñez called "naturally continuous." Three-dimensional space and the plane were thought of as being absolutely continuous, and as the background settings within which geometric objects were located. Space and the plane were assumed to exist independently from geometric figures, which were located "in" or "on" them. Points were seen as being located at particular places in space, or on a plane or line; however, points were not thought of as existing independently of the places where they were located.

At the end of the nineteenth century, after centuries of futile attempts to prove Euclid's Fifth (or Parallel) Postulate, new "non-Euclidean" geometries were developed in which this postulate was violated in various ways. After the emergence of these non-Euclidean geometries, Felix Klein proposed using the concepts of transformations, groups, and invariance to classify and lend order to these new geometries. Thus, in contemporary mathematics since the time of Klein, a "geometry" has been defined as "a study of properties (expressed by postulates, definitions, and theorems) that are left invariant (unchanged) under a group of transformations" (Meserve, 1955, p. 21). For example, the Euclidean transformations preserve the properties of distance, angle measure, area, shape, perpendicularity, parallelism, and incidence, while affine geometry consists of transformations that preserve parallelism.

This redefinition of geometry took place during a time of broader change within mathematics, a change that has been called "the discretization program" (Lakoff & Núñez, 2000, p. 260). The discretization program included the development of analytic geometry and of calculus, and was based on a fundamental metaphorical reconceptualization of numbers. In this reconceptualization, discrete numbers are associated with discrete points on a line (and vice versa), utilizing the number-point conceptual blend described above. Within the discretization program, space was also reconceptualized. Rather than being seen as a naturally and absolutely continuous background or medium within which geometric objects existed, space was now seen as a set of discrete points. This metaphor is summarized in Table 2.1 below. In the table, the source domain, a mathematical set, is utilized to provide redefinitions and inferential structure to the target domain, a geometric space. Thus, for example, under the metaphor, a 2-dimensional space (i.e., the plane) is conceived of simply as a set of points—the notion that a plane is a naturally continuous

Table 2.1. Reconceptualizing Space

A Space is a Set of Points

Source Domain	Target Domain
A Set with Elements	Naturally Continuous Space with Point-Locations
A set	An n-dimensional space—for example, a line, a plane, a 3-dimensional space
Elements are members of the set	Points are locations in the space
Members exist independently of the sets they are members of	Point-locations are inherent to the space they are located in
Two set members are distinct if	Two point-locations are distinct if they are different entities they are different locations
Relations among members of the set	Properties of the space

surface is not a part of this metaphor. Similarly, the properties of the space can be understood as relations among members of the set (relations among point-locations).

Under this metaphor, the objects of traditional geometry are radically reconceptualized:

> There is nothing inherently spatial about a "space." What are called "points" are just elements of the set of any sort.... Like any members of sets, points exist independently of any sets they are in. Spaces, planes and lines— being sets— do not exist independently of the points that constitute them.... A geometrical figure, like a circle or a triangle, is a subset of the points in a space, with certain relations among the points. (Lakoff & Núñez, 2000, p. 263)

This is the basis for the contemporary conceptualization of geometry, and the transformation geometry microworld was built based on this "modern" understanding of geometry. The transformations are to be understood as mappings of the whole plane, that is, functions that map a set of all the discrete points in the plane to themselves. The geometric figures that appear on the computer screen are not meant to be seen as objects separable from the plane, but simply as specific subsets of points. The computer microworld visually highlights these subsets/figures in order to illustrate, graphically, the effects of a given transformation.

Although the microworld may have been designed with a contemporary mathematical perspective on geometry in mind, the students who used the software as a learning environment interpreted it in quite a different way. The learners who worked with the transformation geometry microworld, whether young people or adults, brought to it not the mathematician's commitment to the power of formalism, but instead, a set of intuitions honed through activity and experience in the physical world. In other words, their understanding of geometry was closer to their embodied experience than to the formal definitions of contemporary mathematics. This difference helps to account for the way that the students interpreted and interacted with the computer microworld for transformation geometry, including their "misconceptions."

TRANSFORMATIONS: MOTIONS OR MAPPINGS?

Data on how students interacted with the computer microworld indicate that very different conceptualizations of geometric transformations are held by contemporary mathematicians and by these learners. The students' responses to the computer microworld grew out of an embodied understanding of space (and the plane) as naturally continuous, and as a "place" where objects can "sit" and "move." Their responses and "errors" indicated that when they looked at the graphical window of the micro-world, they were not seeing a mathematician's plane, that is, a set of discrete points, but rather a naturally continuous surface, with objects "on" it. In fact, in some sense, the plane was invisible to them, in that it simply provided a background, *upon which* geometric figures were located. The commands (the geometric transformations) were simply ways that the students could *move* the geometric figures around on top of this plane. By contrast, for the mathematician, the plane consists of an infinite number of discrete points, and transformations are specific, defined mappings of those points that preserve some properties and change others. From the point of view of the discipline of mathematics as it has evolved since the time of Klein and of the discretization program, there are no motions involved in transformations of the plane. Instead, there are mappings of the whole plane, which may be illustrated by highlighting the effects on specific subsets of points.

These two conceptualizations of transformations, the learner's and the mathematician's, are contrasted in Table 2.2. The way that the learners interacted with the computer microworld was consistent with an everyday, naturally continuous conceptualization of space, one that was prevalent within mathematics itself prior to the nineteenth century. On the other hand, the design of the computer environment was consistent with a con-temporary conceptualization of transformations, as mappings *of* the plane rather than motions *on* the plane. This difference in the source domains for understanding transformations can be seen as underpinning the diffi-culties that students had with certain aspects of transformation geometry.

THE ROTATE "BUG" AND EMBODIMENT

The learner's understanding of the transformations in the microworld, as well as the Rotate "Bug" in particular, both arise from directly embodied experience. Our experience of the space we move through is as a

Table 2.2. Two Conceptualizations of Transformations

Learner's Transformations	Mathematician's Transformations
The plane is an empty, invisible background	The plane is a set of points
Geometric figures sit *on* the plane	Geometric figures are subsets of points of the plane
Transformations are physical motions of geo-metric figures on top of the plane	Transformations are mappings of all points of the plane
The "objects" of geometry are points, lines, circles, triangles, etc.	The "objects" of geometry are groups of transformations.

continuous medium, one that is distinct from ourselves and from other objects that exist within it. When we interact with objects on a table (a physical analogue to the mathematician's plane), they sit *on* the table; they are not embedded "subsets" of it. As noted above, the Rotate "bug" is indicative of a difficulty in conceptualizing rotation around a center point that is distant from the L-shape. As human infants, our first physical actions of turning always occur around a center point located within our own bodies, or in the bodies of those physically close to us. Other early experiences with rotation, including the turning of wheels, phonograph records, and the like, are also of this "local" nature. The rotation always occurs around a center that lies within the object itself.

This embodied experience helps to explain why it is difficult for students to understand a rotation (or more precisely, a mapping) of the plane around an arbitrary center point, if this center point is detached from the geometric figure that is the object of their attention. In fact, the most effective method I discovered for clarifying the mathematically "correct" definition of rotation was to ask the student imagine a string between the L-shape and the center point (or to draw a line between the two). This "bridging activity" was both consistent with physical experience with "things that turn", and yet also allowed students to begin to think about rotation about any arbitrary center point. Thus, it was a way to begin to link the students' embodied understanding of rotation with the mathematician's general conceptualization of transformations as arbitrary mappings of the plane.

CONCLUSIONS

In this chapter, students' understandings of transformation geometry, within the context of an exploratory computer environment, are examined as a case of a conceptual tension between an embodied experience of geometry, and a formalized one as utilized in contemporary mathematics. The mathematician sees geometric transformations as mappings of the plane, and the plane is understood as a collection of discrete points, with geometric figures as subsets of points. On the other hand, students understand the plane as something like a naturally continuous surface, and geometric figures as "objects" that sit on top of it. Naturally, then, transformations are seen as motions of objects on the plane, rather than as mappings of the plane to itself. The microworld was constructed to be consistent with the mathematician's view of geometry. However, for the most part, when students interacted with the microworld, there was no visual difference between how the transformations behaved whether seen as motions or as mappings (and, indeed, it was impossible to represent the plane as a collection of distinct points; instead, a "grid" of distinct points was used to allow students to specify the parameters of the transformations). In some cases, however, the microworld behaved in a way that violated students' expectations, for instance, in carrying out a rotation around an arbitrary point on the plane. In this case, the differences in the source domains for knowledge about geometry utilized by the students and by mathematicians leads to different interpretations of rotation. The students will see (or expect to see) a shape that "slides" to the center point, then pivots around it. On the other hand, mathematicians have no trouble in imagining the entire place being mapped around an arbitrary fixed point, while the shape, lengths, and size of the "special" set of points are preserved.

It is certainly the case that one role that computer microworlds can play is to provide students with an arena for reconciling naive understandings and mathematically accepted ones. However, this role can be even more powerful if we are cognizant of the source of learners' initial conceptualizations. Rather than simply seeing these ideas as "misconceptions," we can understand them as, in actuality, conceptualizations that are adaptive and functional outside of the context of formal mathematics, and furthermore, consistent with the way that space and the plane have been thought of within formal mathematics until relatively recently.

The theory of embodied cognition and the analysis of mathematical ideas in terms of fundamental conceptual structures such as metaphors and conceptual blends can provide a powerful perspective on students' learning of mathematics. Rather than seeing students' ideas as deficient versions of correct (and eternally true) mathematical ideas, mathematics educators can utilize this framework to both analyze the development of mathematics, individually and historically, and to design instruction that takes into account and builds on our embodied experiences in the physical world. The re-analysis of data on students' understanding of transformation geometry provides one illustration of this thesis. Additional work from this perspective can not only refine the theory of embodied mathematics but also add to our store of knowledge of how to improve mathematics instruction by connecting with students' embodied understandings.

ACKNOWLEDGMENTS

This paper is based in part on a paper titled "The Nature of Mathematics as Viewed From Cognitive Science," presented at the Third Congress of the European Society for Research in Mathematics Education, Bellaria, Italy (2003).

REFERENCES

Dehaene, S. (1997). *The number sense: How the mind creates mathematics.* Oxford, England: Oxford University Press.

Edwards, L. D. (1989). *Children's learning in a computer microworld for transformation geometry.* Unpublished doctoral dissertation, University of California at Berkeley.

Edwards, L. D. (1997). Exploring the territory before proof: Students' generalizations in a computer microworld for transformation geometry. *International Journal for Computers in Mathematical Learning, 2,* 187–215.

Fauconnier, G., & Turner, M. (2002). *The way we think: Conceptual blending and the mind's hidden complexities.* New York: Basic Books.

Lakoff, G., & Núñez, R. (2000). *Where mathematics comes from: How the embodied mind brings mathematics into being.* New York: Basic Books.

Long, C. T., & De Temple, D. W. (2003). *Mathematical reasoning for elementary teachers.* Boston: Addison Wesley.

Meserve, B. (1955). *Fundamental concepts of geometry.* New York: Dover.

Núñez, R. E., Edwards, L. D., & Matos, J. F. (1999). Embodied cognition as grounding for situatedness and context in mathematics education. *Educational Studies in Mathematics, 39,* 45–65.

Sfard, A. (2000). On reform movement and the limits of mathematical discourse. *Mathematical Thinking and Learning, 2,* 157–189.

CHAPTER 3

SIGNIFYING RELATIVE MOTION

Time, Space and the Semiotics of Cartesian Graphs

Luis Radford

Kant was perhaps the first to have realized how entrenched our knowledge of the world is in the way we experience it through space and time. Since all our acts, even the most mundane, presuppose a temporal and a spatial dimension, space and time, Kant reasoned, constitute the very conditions of knowledge: they are in us and precede all experience whence knowledge results—to use Kant's terminology, space and time are *pure intuitions*. While agreeing with Kant's emphasis on the importance of space and time in the experience we make of the world, current research on epistemology, anthropology, and the arts suggests, however, that space and time are neither apriori conceptual categories, nor the constructs of the allegedly Piagetian universal logico-mathematical structures. Space and time are rather cultural conceptual categories. The culture in which we happen to live not only provides us with the general theoretical framework in which to temporally and spatially

Mathematical Representation at the Interface of Body and Culture, pp. 45–69
Copyright © 2009 by Information Age Publishing

experience our world but also insinuates paths to reflect about it, both at a practical and a theoretical level.

THE ELUSIVENESS OF TIME

Let us go "back in time" for a moment and have a look at a problem historically considered the first in its genre. It is a problem from the eighth century, by Alcuin of York, one of the principal figures of Charlemagne's educational reform. The problem, included in a school textbook—*Problems to Sharpen the Young*—reads as follows:

> There is a field 150 feet long. At one end is a dog, and at the other a hare. The dog chases when the hare runs. The dog travels 9 feet in a jump, while the hare travels 7 feet. How many feet will be traveled by the pursuing dog and the fleeing hare before the hare is seized? (Alcuin, 2005, p. 68)

The problem, written with a didactic purpose, lets us get a glimpse at the manner in which, at this point in the Middle Ages, space and time became object of scientific enquiry and mathematical discourse. The statement of the problem reveals the pregnant phenomenological dimension of a world still not invaded by clocks measuring time with great digital precision. Whereas space is measured by "feet"—an already abstract unit that still keeps its embodied form and evokes the spatial relationship between the motion of an individual and its surrounding—time is not explicitly mentioned in the problem. To compare the space traveled by the dog and the hare, Alcuin resorts to the idea of *jump*. In a jump, the dog travels 9 ft, while the hare travels 7 ft. How then, without explicitly employing the idea of time, can this problem be solved?

Let us turn to the solution. Alcuin says:

> The length of the field is 150 feet. Take half of 150, which is 75. The dog goes 9 feet in a jump. 75 times 9 is 675; this is the number of feet the pursuing dog runs before he seizes the hare in his grasping teeth. Because in a jump the hare goes 7 feet, multiply 75 by 7, obtaining 525. This is the number of feet the fleeing hare travels before it is caught. (Alcuin, 2005, p. 68)

The first calculation (i.e., the half of 150) corresponds to the number of jumps. The question is: are these the dog's jumps or the hare's jumps? Jump (*saltu* in the original Latin), is, like foot, an abstract idea: it is neither the dog's nor the hare's. It evokes a phenomenological action that unfolds over a certain *duration*. It is only in this oblique way that time appears in the problem. After each jump, the dog comes 2 ft closer to the hare. Thus, the dog will need 75 jumps to catch the hare. This number of jumps is multiplied by the 9 ft that the dog goes in a jump—the medieval expression of "speed"—and then by 7, that is, the number of feet that the hare goes in a jump. The resulting numbers are the space travelled by each animal.

Problems like this became popular later on. There is a problem in a fourteenth century Italian manuscript, composed by Piero dell'Abacco that reads as follows:

> A fox is 40 paces ahead of a dog, and three paces of the latter are 5 paces of the former. I ask in how many paces the dog will reach the fox. (dell'Abacco as translated by Arrighi, 1964, p. 78)

Here, distance is measured by "paces" and, like in Alcuin's problem, time is mentioned implicitly through *motion*.

The difficulties in dealing with time are well known when it comes to philosophers and epistemologists. Like Alcuin and dell'Abacco, Aristotle related time to motion. As he said, "we perceive movement and time together ... time and movement always correspond with each other" (Aristotle, Physics, IV, XI, p. 62). And Grize (2005), commenting on the elusiveness of time as expressed in the works of Aristotle and Bishop Augustine, notes that within this perceptual frame of reference, "A distance is measured in relation to a length; time, however, cannot be measured by a time anymore than a temperature can be measured by a temperature unit" (p. 69). Time, hence, remained an implicit notion, embedded in the duration of motion (like the sun's analogical projected shadow on the sundials or the bodily jumps Alcuin's text talks about), until mechanical clocks, exploiting a rhythmic pendulous repetition, extricated it from its conceptual limbo and made time a precise theoretical object in its own right.

It therefore does not come as a surprise that in many Medieval and early Renaissance mathematical problems that were accompanied by drawings, time remains expressed in the perceptual motion of the moving objects (see Figure 3.1). Like Alcuin, dell'Abacco was also a teacher. We may conjecture that in discussing these problems with their students, both schematically enacted the moving objects through gestures or body movements.

Be this as it may, it might not be useless to compare the medieval and modern solutions of the previous problems. Let us consider dell'Abacco's problem. The medieval solution proceeds by a comparison of traveled distance: three dog' steps are equal to 5 fox steps or 3D = 5F. Although dell'Abacco does not mention it explicitly, he assumes that *while the fox goes one step the dog goes one step as well*. Time appears in the problem through this hidden assumption. From 3D = 5F, using a rule of three,

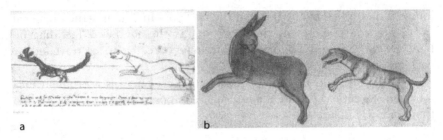

Figure 3.1.   a. The drawing accompanying Alcuin's problem. b. The drawing accompanying dell'Abacco's problem.

dell'Abacco deduces that 5D = 8 1/3F. Thus, when the dog takes 5 dog steps, the fox takes 5 fox steps and the distance between the fox and the dog diminishes by 3 1/3 fox steps. Knowing that they are 40 fox steps apart, and continuing using the rule of three, dell'Abacco concludes that the dog will need to go 60 dog steps to reach the fox.

The modern reader might find this fourteenth century solution a bit strange, if not unclear. A modern solution starts, in fact, by expressing the problem in terms of velocities, hence in relating traveled distance to units of time. But to think in terms of specific quantified "small" units of time (like seconds) is exactly what the medieval thinkers did not do. In tune with the modern solution, let us assume that the fox travels 5 fox steps per unit of time; then, the dog travels 5 dog steps or 25/3 fox steps per unit of time. Referring to the spatial place where the *dog* was located at the beginning of its race, the distance traveled by the dog (expressed in fox steps) is:

$$d_d = 8\tfrac{1}{3}t \qquad (3.1)$$

Referring to the *previous* spatial point (i.e., the *dog*'s initial place), the distance travelled by the *fox* is:

$$d_f = 40 + 5t \qquad (3.2)$$

The point at which the dog catches the fox is characterized by the equality:

$$8\tfrac{1}{3}t = 40 + 5t \qquad (3.3)$$

And when we solve this equation, what we get is not the distance travelled by one or the other, but *time*—the time that both animals have been in motion:

$$t = \frac{40}{3\frac{1}{3}} \text{ or } t = 12 \tag{3.4}$$

The sense of our modern solution is quite different from Alcuin's and dell'Abacco's. We, more or less consciously, forget what $8\frac{1}{3}$ and $5t$ mean and mechanically subtract them to get $3\frac{1}{3}t$, which we then equate to 40. From there, we find t, the numerical value of this elusive concept that was not even mentioned in either dell'Abacco's statement of the problem or in its solution. Now we substitute the value of t in the first equation and get

$$d_d = 8\frac{1}{3} \times 12 = 100 \text{ fox steps (or 60 dog steps).}[1]$$

UNIFYING SYSTEMS OF KNOWLEDGE REPRESENTATION

There is also something very different in the modern solution. The motion of both the dog and the fox were referring to a *same spatial point* (the initial position of the dog in the race). In Medieval and early Renaissance problems, the motion of the moving objects involved remained without being described into a unifying system of reference (see dell'Abacco's solution to the fox problem in note 1, where calculations are made by *comparison* of "speeds" and not by *integration* of data into a same totality). From a semiotic point of view, there is a striking similarity between the mathematical problems and the paintings and drawings of the Middle Ages and early Renaissance. Their respective signs revolve around a main "subject" without being linked by a truly functional unifying system of representation. For instance, objects surround the main subject (e.g., the saint) in a juxtaposed manner. Each object contributes to the whole meaning of the drawing or painting by addition of its particular meaning (see Figure 3.2a).

The order of the signs in paintings and mathematical texts became profoundly transformed by the concurrent invention of the technique of perspective and algebraic symbolism (Radford, 2006). The new cultural forms of knowledge representation continued to privilege a certain subject, but now there was a relational link ensuring the relationship between a chosen central object and other objects (see Figure 3.2b, c).

The emergence of a Cartesian system of coordinates and its central point $(0, 0)$ was one of the most sophisticated ways in which to express the complex set of relations between the objects described in the situation at hand. This was one of the crucial developmental steps in the mathematical

Figure 3.2. a. A drawing from the 5th century Vat. Lat. 3867 manuscript (Brown, 1990, p. 19); objects are juxtaposed. b. Uccello's painting *The Miracle of the Desecrated Host*, Scene 1 (ca. 1468); objects are related in a unifying perspective system. c. Contemporary formulas of the distance in an integrated system where reference to a same spatial point and common time become the explicit organizing element..

study of motion. The creation of such a system required, in particular, a mathematical reconceptualization of space and time. It is hence not surprising that Cartesian graphs rest on a sophisticated manner of signifying that, from an ontogenetic viewpoint, is far from transparent to novice students.

LEARNING AND THE HISTORICAL DIMENSION OF KNOWLEDGE

What is it then that we ask the students to accomplish when we expect from them to describe problems about motion through algebraic formulas and Cartesian graphs? As intimated by my previous remarks, a graph is a complex mathematical sign. It serves to depict, in specific ways, certain states of affairs. Instead of being merely a reproduction of these, a graph supposes a selection of elements: what it depicts is *relationships* between them. This is why the making of a graph of an elementary phenomenon, such as the motion of an object, is like putting a piece of the world on paper (or electronic medium). But because Cartesian graphs are not copies of the phenomena that they depict or represent, making and interpreting them is not a trivial endeavor. A Cartesian graph rests on a sophisticated syntax and a complex manner of conveying meanings. It is the understanding and creative use of this complex historically constituted cultural form of signification that we expect the students to accomplish when dealing with graphs, and, as we know, it does not go without justified difficulties.

The investigation of the difficulties surrounding students' understanding of graphs has been an active research area in mathematics education.[2] In this chapter, I contribute to the research on graphs by looking at students' processes of graph understanding. I am interested in particular in researching the way in which students attempt to make sense of graphs related to problems of relative motion—an area little investigated thus far. To so do, drawing on a Vygotsky's historical-cultural school of thought, I consider here mathematical thinking as a cultural and historically constituted form of reflection and action, embedded in social praxes and mediated by language, signs and artifacts (Radford, 2006). A Cartesian graph is an artifact for dealing with and thinking of cultural realities in a mathematical manner. But, as mentioned previously, this artifact is not transparent: it bears the imprint and sediments of the cognitive activity of previous generations which have become compressed into very dense meanings that students have to "unpack," so to speak, through their personal meanings and deeds.

This process of "unpacking" is the socially and culturally subjective situated encounter of a unique and specific student with a historical conceptual object—something that I have previously termed *objectification*

(Radford, 2003). The construct of *objectification* refers to an active, creative, imaginative, and interpretative social process of gradually
becoming aware of something. Within this context, understanding the
making and meaning of a graph, the way it conveys information, the
potentialities it carries for enriching and acting upon our world, rests on
processes of objectification mediated by one's voice, others' voices and
historical voices (Boero, Pedemonte, & Robotti, 1997). Objectification is
indeed a multi-voiced encounter between an "I," an "Other," and (historical and new) "Knowledge."

Now, one of the distinctive traits of human cognition is its *multimodal*
nature. What this means is that thinking is profoundly dependent upon
the cultural artifacts that we use and our own body. As Gallese and Lakoff
(2005) expressed the idea, "the sensory-motor system not only provides
structure to conceptual content, but also characterizes the semantic content of concepts in terms of the way that we function with our bodies in
the world" (p. 456). In other words, signs (language included), artifacts,
and our body along with its various senses are vehicles for thought. Within
this context, in the objectification of mathematical knowledge, recourse is
made to body (e.g. kinesthetic actions, gestures), signs (e.g., mathematical
symbols, graphs, written and spoken words), and artifacts of different
sorts (rulers, calculators and so on). In the practical investigation of students' understanding of graphs, I therefore pay attention to the students'
gestural, kinesthetic, symbolic and discursive activity as they attempt to
make sense of a graph.

MAKING SENSE OF GRAPHS—A CLASSROOM EXAMPLE

In this chapter, I discuss the attempt made by one group of tenth-grade
students in their effort to understand a graph representing the relative
motion of two bodies. The data, which comes from a five-year longitudinal research program, was collected during classroom lessons that are
part of the regular school mathematics program in a French-language
school in Ontario. In these lessons, designed by the teacher and our
research team, the students spend substantial periods of time working
together in small groups of 3 or 4. At some points, the teacher (who interacts continuously with the different groups during the small group-work
phase) conducts general discussions allowing the students to expose, compare and improve their different solutions.

The data that will be discussed here comes from a lesson featuring a
graphic calculator TI 83+ and a probe——a Calculator-Based Ranger or
CBR (a wave sending-receiving mechanism that measures the distance
between itself and a target). The students were already familiar with the
calculator graph environment and the CBR. In previous activities, they

had dealt with a fixed CBR and one moving object. In the activity that I discuss here, the students were provided with a graph and a story. The graph showed the relationship between the elapsed time (horizontal axis) and the distance between two moving children (vertical axis) as measured by the CBR (see Figure 3.3). The students had to suggest interpretations for the graph and, in the second part of the lesson (not reported here), to test it using the CBR.

Here is the story:

> Two students, Pierre and Marthe, are one meter away from each other. They start walking in a straight line. Marthe walks behind Pierre and carries a calculator plugged into a CBR. We know that their walk lasted 7 seconds. The graph obtained from the calculator and the CBR is reproduced below.

(See Figure 3.3 for the illustration and the graph.) The disposition of the axes in the Cartesian graph reflects the modern concept of space and time as continuous variables represented by oriented lines. We chose three main "events" to be interpreted—the segments AB, BC and CD. They were different not only in their successive positions in the graph but also in their orientation. Within the Cartesian semiotic system, "events" signify in a *relational* and *unifying* manner. Of course, as already mentioned, this historically constituted manner of signifying is far from trivial. In what follows, we see that in order to be able to interpret these events, the students will have to unpack space and time from the phenomenological expression embedded in motion.

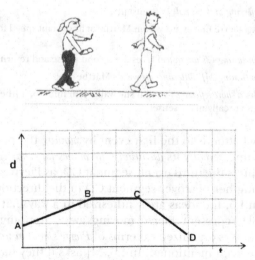

Figure 3.3. The story illustration and the graph given to the students.

STUDENTS' FIRST INTERPRETATION

I focus on one 3-student group and present some excerpts of the students' processes of interpretation, with commentaries on the progressive manner in which objectification was accomplished. The students were Maribel (M), Marie-Jeanne (MJ) and Carla (C). After discussing the problem for a few seconds, in Line 1 (L1), Maribel offers an interpretation of segment AB:

1.   M: Then, there (*she moves the pen along the segment AB*) he moves for what ... 3 seconds?

2.   MJ: Yeah! ...

3.   M: Then here (*referring to segment CD*) he ...

4   MJ: (*Moving the pen over segment CD, from D to C*) He goes backwards for 2 seconds ... (*See Figure 3.4a*)

5.   M: (*Summarizing the discussion,* she *moves the pen on the desk and says*) He moves away from Marthe for 3 seconds, and then he stops (*she stops the pen on the desk at a point that would correspond to the point B in the graph; see Figure 3.4b*), so (*referring to segment BC, she moves the pen further along the desk; see Figure 3.4c*) he might have like ... dropped something for 2 seconds, and (*moving the pen back this time; Figure 3.4d*) he returns towards Marthe ...

6.   MJ: Well ... does it (*referring with a pointing gesture to the distance axis*) have to have like ... specific things?

7.   C: His speed increases a bit ...

8.   MJ: ... What I mean, it's, like, the distance, does it have to be specific?

9.   M: No.

10.   C: ... If the speed increases ... it would be a curve, right?

11.   MJ: (*Referring to the speed*) It's constant.

12.   C: So ... Pierre moves away from Marthe at a constant speed for 3 seconds ...

13.   MJ: (*Continuing C's utterance*) takes a 2 second pause and returns ...

14.   M: (*Continuing MJ's utterance*) towards Marthe...

15.   MJ: (*After a short reflection*) Well, if she walks with him, so, it [the graph] doesn't really make sense!

In L1, Maribel attends to the first event by *moving* the pencil from A to B while mentioning *verbally* its *duration* ("*he moves for what ... 3 seconds?*"). In L4, MJ interprets the third event, segment CD, as Pierre going back for 2 seconds, moving her pen over segment CD in the direction from D to C (Figure 3.4a). In L5, the ideas are synthesized in a way that the segments AB, BC and CD represent Pierre moving away, stopping and coming back. The synthesis is organized in terms of *Pierre's motion* and its *duration*. Distance has not been mentioned. In L6, MJ asks if they have to consider particular values for the distance. However, the focus is put on a vague

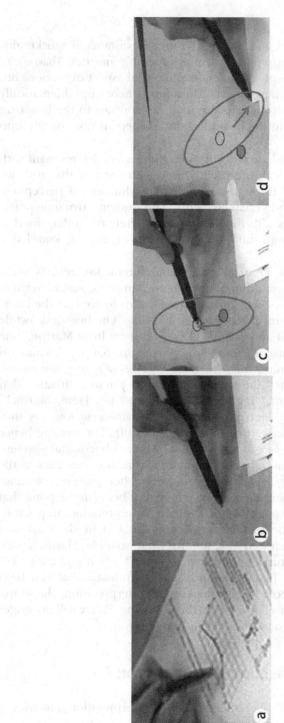

Figure 3.4. Some gestures made by MJ (a) and M (b–d) during the first interpretation of the graph.

qualitative idea of speed and, in L9, the idea of distance is quickly dismissed. The students' approach resembles Alcuin's and dell'Abacco's in one point: the emphasis on the phenomenological aspect of motion. But it differs in other important aspects. Duration is here unproblematically quantified in terms of seconds; furthermore, in contrast to the historical texts discussed previously, the quantification of speed is not brought into the students' discourse.

Although the students' current interpretation is not yet resonant with the expected mathematical interpretation, we can see that the students' ideas have been forged through a complex coordination of perceptual, kinesthetic, symbolic, and verbal elements. The students' dynamic pointing gestures and actions with the pen are not merely redundant mechanisms of communication, but key embodied means of knowledge objectification.

A closer look at the gestural and verbal interaction reveals some aspects of the students' unfolding interpretation. In particular, in her synthesis in L5, Maribel's gestures and actions allow us to see that the interpretation is entangled in a *juxtaposition of spaces*. On her desk, while referring to segment AB and saying "He moves away from Marthe," she moves the pen as if enacting Pierre's walk (Figure 3.4b). This motion occurs in what we may term the *phenomenological space of imagined motion*. While she says, "he stops," she *continues moving the pen* in a direction that now evokes the horizontal segment BC (Figure 3.4c). Here, Maribel's motion is not enacting Pierre's walk, but the *passing of time* (as they explain, in this part, Pierre is considered to be still). This gesture hence occurs in the *Cartesian space of representation*, where a horizontal segment represents the passing of time. And right after this, she goes back to the *phenomenological space of imagined motion*, where her gesture continues evoking Pierre's walk: Now Pierre is imagined as if being at the point that in his walk would correspond, in the students' interpretation, to point C. Instead of following the inclination of segment CD in the Cartesian graph, Maribel moves her pen *back*, towards what would be Marthe's position, saying, "and he returns towards her" (Figure 3.4d). It is at the end of this episode that Marie-Jeanne reminds her group-mates that Marthe is moving too, so that, according to the current interpretation, the graph "doesn't really make sense!" For, if Marthe is moving, Pierre will no longer find her when he walks back towards her initial position!

A SECOND INTERPRETATION

Twenty seconds later, Maribel offers a refined interpretation that tries to address the issue raised by Marie-Jeanne:

16.   M: Well technically, he walks faster than Marthe ... right?

17.   MJ: She walks with him, so it could be that [...] She is walking with him, so he can walk faster than her (*she moves the pen on segment AB; see Figure 3.5*). [He] stops (*pointing to points B and C*) ...

18.   M: No, there (*referring to the points B and C*) they are at the same distance ...

19.   C: (*After a silent pause, she says with disappointment*) Aaaaah!

The graph interpretation has changed: In the previous episode, the segments were seen as predicating something about *Pierre*. Marthe was not really part of the story told by the graph. In L16, Maribel introduces the two-variable comparative expression "X walks faster than Y." In L17, Marie-Jeanne reformulates Maribel's idea in her own words, while producing a more sophisticated interpretation. Indeed, L17 contains three ideas: (a) Marthe walks with Pierre; (b) Pierre walks faster than her, and (c) Pierre stops.

Although improved, the interpretation, as the students realize, is not free of contradictions. These contradictions result in part from incautiously endowing the segments with meanings coming from the *phenomenological space of imagined motion* and the *Cartesian space of representation*. The meanings overlap, resulting in a global incoherent interpretation. Even if, at the discursive level, Marthe is said to be walking (L17), segment AB is still understood as referring to Pierre's motion (MJ says at the end of the movement of his pen in Figure 3.4c, "[He] stops"). However, segment BC is interpreted not in terms of *motion* but of *distance* (L18): BC is interpreted as indicating that the distance between Pierre *and* Marthe remains the same during this period of time. So, while segment AB is about Pierre's motion, segment BC predicates something about both children's distance. The interpretation of the events does not yet fit into a unifying systemic logic of

Figure 3.5.   MJ moves the pen from A to B, meaning Pierre's motion (L17).

knowledge representation. The oddity of the interpretation leads to a tension that is voiced by Carla in Line 17 with an agonizing "Aaaaah!" The partial objectification bears an untenable incongruity.

THE TEACHER

The students continued discussing and arrived at a new interpretation: In the story-problem, the students are told that Pierre and Marthe are 1 meter apart from each other. Thus, the new interpretation: Pierre and Marthe maintained a distance of 1 meter apart throughout. The students could not agree on whether or not this interpretation was better than, or even compatible with, Maribel's interpretation (L16). Having reached an impasse, the students decided to call the teacher (T). When he arrived, Marie-Jeanne explained her idea, followed by Maribel's opposition; it is this opposition that is expressed in L20:

20. M: No, like this (*moving the pen along segment AB*) would explain why like, he goes faster, so it could be that he walks faster than her ...

21. T: Then if one is walking faster than the other, will the distance between them always be the same?

22. M: No, (*while moving the pen along AB, she says*) so he moves away from the CBR and then.... What happens here (*pointing to segment BC*), like?

23. MJ: He takes a brake.

24. T: So, is the CBR also moving?

25. M: Yes.

In L21, the teacher rephrases the first part of Maribel's utterance (L20) in a hypothetical form to conclude that, under the assumption

Figure 3.6.   The teacher moves the pen back and forth between the intersection of the axes and point A.

that Pierre goes faster, the distance cannot be constant. Although inconclusive from a logical point of view, the teacher's strategy helps move the students' discourse to a new conceptual level. Maribel's L22 utterance shows, indeed, that the focus is no longer on relative speed but on an emergent idea of *relative distance*. In L22, the moving gesture along segment AB is the *same* as Marie-Jeanne's in Figure 3.5, but its *content* is *different*: although the gesture still enacts Pierre's motion in the *phenomenological space of imagined motion*, it signifies Pierre *moving away* from the CBR. However, as shown in L22, the students still have difficulties providing a coherent global interpretation of the graph. How to interpret BC within the new relative motion context? Drawing on Maribel's utterance (L22), in L24 the teacher suggests a link between Marthe and the CBR, but the idea does not pay off as expected. He then tries something different.

THE MEANING OF SEGMENT 0A

26.   T: OK. A question that might help you.... A here (*he circles point A*). What does A represent on the graph? (*He moves the pen several times between the intersection of the axes and A; see Picture 6*)

27.   MJ: Marthe.

22.   M: No, (*while moving the pen along AB, she says*) so he moves away from the CBR and then.... What happens here (*pointing to segment BC*), like?

28.   T: This here (*pointing to the intersection of the axes*), is it zero? (*he writes 0 at the intersection of the axes*)? We'll only talk about the distance. OK? (*He moves the pen again over 0A as in Line 26*).

29.   MJ: 1 meter.

30.   T: (*Rephrasing MJ's answer*) it (*i.e., the segment 0A*) represents 1 meter, right? ... 1 meter in relation to what?

31.   M: The CBR ...

32.   T: OK. So, does it represent the distance between the two people?

33.   M: So this (*moving the pen along the segments*) would be Pierre's movement and the CBR is 0.

34.   MJ: (*Interrupting*) First he moves more ...

Capitalizing on the emerging idea of relative distance, the teacher's strategy now becomes one of calling the students' attention to the *relational meaning* of a particular segment—the segment OA. This segment has passed unnoticed so far. It refers, in relational terms, to the beginning of the story-problem—the distance between Pierre and Marthe. In this sense, it appears as a good point from which to launch a Cartesian interpretation of the graph in general and segment AB in particular.

The teachers captures the students' attention in three related ways:

- writing (by writing 0 and circling point A);
- gesturing (by moving the pen back and forth between 0 and A);
- and verbally (L26).

Since the students see the graph as a kind of map of the position of Pierre and Marthe, they insist on locating them somewhere in the Cartesian graph. Thus, in L27, point A is associated with Marthe. However, it is precisely not this phenomenological reading that the teacher aims at, but a relational one. So, in L28, he formulates the question in a more accurate way and takes advantage of the answer to further emphasize (L30) the idea of the relative meaning of the distance. Line 33 includes the awareness that the CBR has to be taken into account, while L34 is the beginning of an attempt at incorporating the new significations into a more comprehensive account of the meaning of the graph.

The teacher left the students saying, "I do not say anything more!" and went to talk to another group. The students thus entered into a new phase of knowledge objectification. They continued discussing in an intense way. Here is an excerpt:

35.   C: He moves away from her, he stops then comes closer.

36.   M: But she follows him…. So, he goes faster than she does, after, they keep the same distance apart.

In L35, Carla still advocates an interpretation of the graph that suggests a fragile understanding of relative motions. In the first part, she makes explicit reference to Marthe ("He moves away from her"), but in the second and third part of the utterance, Marthe remains implicit. The *phenomenological space of imagined motion* and the *Cartesian space of representation* are not linked suitably yet. As a result, an ambiguity remains.

In L36, Maribel offers an explanation that seems to overcome the ambiguity. Even though the segment AB is expressed in terms of rapidity, the previously reached awareness of the effect of rapidity in the increment of distance makes the interpretation of BC coherent. Maribel recapitulates the students' efforts and says before the group starts writing an interpretation:

Maybe he [Pierre] was at 1 meter (*pointing to A*) and then he went faster; so now he is at a distance of 2 meters (*moving the pen in a vertical direction from BC to a point on the time axis, see Figure 3.7a, b*); and then they were constant and then (*referring to CD*) they slowed down. Would that make sense?

For the first time in their process of objectification, the students provide an interpretation that stresses the relational meanings of the Cartesian space of representation. The interpretation still needs to be refined. For instance, in the interpretation of CD, Maribel did not specify in which manner they slowed down. Was it Pierre who slowed down? Was the reduction of distance the effect of Marthe increasing her speed? Was it something else? Nonetheless, the students were able, to a certain extent, to put into correspondence the relational meaning and the phenomenological space of imagined motion. Key in this accomplishment was Maribel's vertical gesture (Figure 3.7), which makes clear the explicit insertion of the idea of distance in the students' discourse. To understand the students' process of objectification, this gesture needs to be put into correspondence with the teacher's gesture shown in Figure 3.6.

Literary critic Mikhail Bakhtin (1986) once remarked that "Each word contains voices that are sometimes infinitely distant, unnamed, almost impersonal (voices of lexical shadings, of styles, and so forth), almost undetectable, and voices resounding nearby and simultaneously" (p. 124). Figures 3.6 and 3.7 suggest that the same is true of gestures: Maribel's gesture contains the conceptual intention of the teacher's gesture. Naturally, Maribel's gesture is not just a copy of the teacher's; it has been endowed with personal tones and displayed in a different part of the graph. Nonetheless, it bears an almost undetectable voice that has served as inspiration for seeing something new.

In writing their answer, the students, however, realized that something important was missing: the interpretation needed to include Marthe in an explicit way. Naturally, writing requires one to make explicit, and thereby objectify, relationships that may remain implicit at the level of speech and gestures. Maribel's activity sheet contains the following answer:

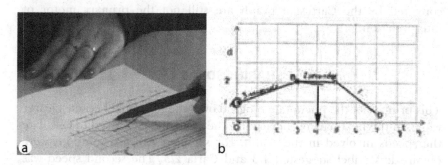

Figure 3.7.   a. Maribel makes a vertical gesture that goes from BC to the time axis and is indicated by an arrow in (b). This gesture is a generalization of the teacher's gesture (Figure 3.6).

Pierre moves away from Marthe by walking faster for 3 seconds. He is now 2 meters away from her. They walk at the same speed for two seconds. Pierre slows down for two seconds so he gets closer to Marthe. (Maribel)

The text is now about the relative distance of Pierre and Marthe (compare this text to Maribel's interpretation in the previous section). It is organized in accordance with the sequentiality of a *common* time—an element whose importance was pointed out in the second section and in Figure 3.2. The first sentence tells us how to imagine what happened during the first three seconds. The emphasis is on the distance: As a result of walking faster, Pierre is "now" 2 meters away from Marthe. The interpretation of segment BC follows: Pierre and Marthe are said to be walking for 2 seconds, at the same speed. Here, the students do not feel the need to tell us that the distance between them remains the same. What was hard to figure out was indeed that here they were walking at the same speed. Thus, this is what needs to be said. In addition, as in the first and last sentence, speed is left without being quantified. In the third part, the idea of the distance is brought to the fore again: Pierre slows down and so he gets closer to Marthe.

Here we can see that the link between the *phenomenological space of imagined motion* and the Cartesian space of representation was improved. The static segments of the Cartesian graph were endowed with a dynamic interpretation where relational aspects of relative distance became partially linked to the phenomenological space of imagined motion. The central concept in the production of the students' narrative was motion. Moving in a certain way (faster, at the same speed, slowing down) explained what happened with the distance between Pierre and Marthe. We may say that the students' interpretation remains primarily phenomenological, rather than relational. In other words, the relational meanings conveyed by the Cartesian graph are still not the primary motor of interpretation.

SHARING IDEAS

This primacy of the phenomenological over the relational became clearer when, in the next part of the activity, the students were asked to calculate the speeds involved in the graph. For segment AB, two solutions were obtained: Maribel suggested 1/3 and Carla 2/3. The second speed was based on the idea that Marthe was at the origin of the graph. Following a classroom practice encouraged by the teacher—where students are invited to visit other groups to submit, compare and discuss their ideas—Marc

(Mc), a student from a group at the opposite side of the classroom, came to discuss with Carla's group.

Maribel explains her calculations to Marc:

37.   M: This (*referring to the speed associated with the first event—segment AB*) is one third.... Do you understand?

38.   C: (*Opposing Maribel's idea!*) No!

39.   M: (*Pointing to the origin of the Cartesian graph; see Figure 3.8a*) Marthe, isn't she here?

40.   Mc: No, Marthe is (*trying to point somewhere, but his finger never lands on the graph*).... Okay. (*He abandons the idea of locating Marthe in the Cartesian graph and starts a different line of thought*). They both begin at 1 meter (*pointing with the back of the pen to segment OA; see Figure 3.8b*) ...

41.   No, she starts here (*pointing to the Cartesian origin*) ...

42.   'Kay. There is 1 meter here (*pointing to OA; Figure 3.8c*). This is the distance between both people (*he draws an arrow between Pierre and Marthe to signify the initial distance between them in the drawing accompanying the story-problem; Figure 3.8d*).

43. M & C: (*at the same time*) Yeah.

44.   Mc: Here (*he draws an arrow in from of Pierre*) he moves forward faster or (*he draws an arrow behind Marthe*) she moves more slowly. So then, it makes a difference in the distance (*Figure 3.8e*).

45.   M: Yes.

46.   Mc: The line increases (*moving the pen along segment AB; the arrow in Picture 6 indicates the sense of the gesture*) because you have more than one [unit of] distance. Ok? Here (*pointing to segment BC*), it's a straight line ... they are moving at the same speed ...

47.   C: Yeah.

48.   Mc: Here (referring to segment CD), she moves forward faster or he slows down so that the distance is smaller.

49.   C: Why does the distance go down again?

50.   Mc: Because they are closer. So the distance between the guy and the CBR ... is smaller.

Marc displays an ability to move between the phenomenological and relational spaces. He shows a clear understanding of how these two spaces relate to each other. The length of the Cartesian segment OA is translated into the phenomenological space (L40 and L42). In L46, the increase of distance in the phenomenological space is related to the inclination of segment AB.

This discussion can be seen as occurring in a zone of proximal development created by the students within the spirit of the circulation of ideas

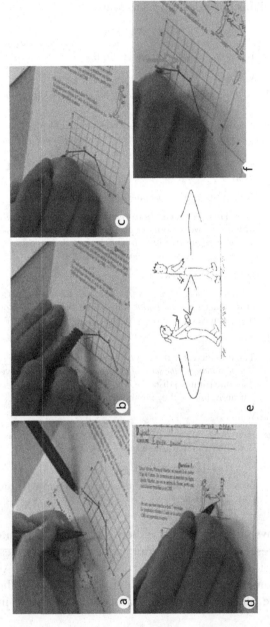

Figure 3.8. Marc's gestural activity during the discussion of the graph. Marc links the phenomenological space of imagined motion and the Cartesian graph in a clear way.

in the classroom encouraged by the teacher. The zone of proximal development allowed Marc to better understand that the graph is not really about locating Marthe somewhere in it: the graph is rather about relative distances. The zone of proximal development helped Carla's group to enhance their understanding of the graph and the complex historically formed cultural logic behind the Cartesian graph.

SYNTHESIS AND CONCLUDING REMARKS

In the first part of this chapter, I discuss two historical problems about the meeting point of two objects moving at different speeds. As pointed out, the solution involved calculations made by *comparison* of "speeds" and not by the *integration* of data into a same totality. Our brief historical excursion allowed us to remark that the concept of time remained rather implicit in the formulation and the solution of these types of problems.[3] It may be true, as Koyré (1966) notes in his studies on pre-modern scientific thought, that, in problems about motion, it was more difficult to think in terms of time than in terms of space. Time appeared imbricated in the concept of motion and could only be extracted from it at great pains.[4] Galileo's own account is eloquent. Commenting on his experiment on an inclined plane, he says:

> As to the measure of time, we had a large pail filled with water and fastened from above, which had a slender tube affixed to its bottom through which a narrow thread of water ran; this was received in a little beaker during the entire time that the ball descended along the channel [carved on the inclined plane] or parts of it. The little amounts of water collected in this way were weighed from time to time on a delicate balance, the differences and ratios of the weights giving us the differences and ratios of the times, and with such precision that, as I have said, these operations repeated time and again never differed by any notable amount. (Galileo, 1638, p. 170)

The new needs brought about by the cultural and economical contexts of the Renaissance led to a reconceptualization of space and time and the emergence of new unifying systems of artistic and scientific knowledge representation (Figure 3.2). One of the most sophisticated examples of such systems is the Cartesian plane, whose laborious constitution required centuries of progressive refinements. In motion problems, the Cartesian plane entails the description of events in terms of *common* spatial and temporal points of reference. The spatial-temporal location of *all* objects is described in relation to these distinguished referential points.

The Cartesian plane allows one to grasp *visually* the evolution of a phenomenon. For Alcuin the problem was not to determine the remaining distance between the dog and the hare after each jump, but the point at which the former catches the latter. He might have found it very curious that one could be interested in calculating the remaining distance for each value of time. Indeed, the interest in following with minute detail the evolution of phenomena became important when attention started to be given to problems of *variation* in the eighteenth and nineteenth centuries.

Now, the systemic and formal structure of a Cartesian plane affords the representation of other more general phenomena—like equipotential curves (see Roth, 2003) or relative motions. In the case of relative motions, what is signified by the distance axis is the relative distance between the moving objects. An important level of indeterminacy is introduced: It is not possible to tell where the objects are in the phenomenological world, for what is known is only the distance between them. This historically constituted form of representing knowledge is far from evident for the novice students. The Cartesian graph bears the sediments of previous generations of cognitive activity and understanding its mode of signifying is, for the students, the outcome of a lengthy process of unpacking knowledge that is termed here *objectification*. Objectification, in fact, is a social process related to the manner in which students become progressively aware, through personal deeds and interpretations, of the cultural logic of mathematical entities—in this case, the complex mathematical meanings that lie at the base of the ways in which Cartesian graphs are used to describe some phenomena and convey meanings.

The data I present here suggest that one of the most important difficulties in understanding the graph was: (a) overcoming an interpretation based on a phenomenological reading of the segments and their descriptions in relation to a fixed spatial point, and (b) the attainment of an interpretation that puts emphasis on relative relations. The question is not to "forget" the phenomenological realm. It is rather to link, in a suitable way, the phenomenological space of imagined motion with the Cartesian space. The logic of interpreting a Cartesian representation of relative motion became progressively apparent for the students through an intense activity mediated by multiple voices, gestures, and mathematical signs. Crucial to this endeavor was the teacher's intervention and the group's discussion with Marc. The teacher was indeed able to call the students' attention to the relationship between segment 0A and the initial distance between Pierre and Marthe, thereby creating some conditions for the evolution of meanings both at the discursive and gestural levels. The teacher's coordination of words with the sequence of similar gestures and signs in the Cartesian graph (Figure 3.6) helped the students understand the meaning of segment 0A in the context of the problem. Segment 0A entered the universe of discourse and gesture, and its length started being considered as the distance between Pierre and Marthe at the beginning of their walk. Without teaching the meaning directly, the teacher's interactional analysis of the meaning of segment 0A was understood and generalized by the students in a creative way (Figure 3.7). After the teacher's intervention, the students' gestures became more and more refined as did their words: whereas their first gestures were about Pierre's motion, their last gestures were related to distances in a meaningful relational way.

The evolution of meanings was deepened during the discussion with Marc, who also insisted on the meaning of 0A, but went further, offering a way to relate the other segments of the Cartesian graph with Pierre and Marthe's walk in a manner that was consonant with what the students had accomplished by themselves. Borrowing a term from Bakhtin (1981), I want to call the transformative objectifying process *heteroglossic*, in that heteroglossia, as I intend the term here, refers to a locus where differing views and forces first collide, but under the auspices of one or more voices (the teacher's or those of other students), are then momentarily resolved at a new cultural-conceptual level, nonetheless awaiting new forms of divergence and resistance.

ACKNOWLEDGMENTS

This article is a result of a research program funded by The Social Sciences and Humanities Research Council of Canada/Le Conseil de recherches en sciences humaines du Canada (SSHRC/CRSH). A previous version of it appeared in Miranda, Radford, and Guzmán (2007).This article is a result of a research program funded by The Social Sciences and Humanities Research Council of Canada/ e Conseil de recherches en sciences humaines du Canada (SSHRC/CRSH). A previous version of it appeared in Miranda, Radford, and Guzmán (2007). Figure 3.1.a comes from ms. Magl. XI, 86, fol. 28r and is reproduced here by kind permission of the Ministero per i Beni e le Attività Culturali della Repubblica Italiana/ Biblioteca Nazionale Centrale di Firenze. Figure 3.1.b, *Il cane e la lepre*, comes from Alcuino di York. Giochimatematici alla corte di Carlomagno (a cura di Raffaella Franci), and is reproduce here by kind permission of Le Edizioni ETS, Pisa, Italy. Figure 3.2.a comes from Vat. Lat. 3867, Vergilius Romanus (fifth century), Ecloga II, 1-4, f. 3v. and is reproduce here by kind permission of Biblioteca Apostolica Vaticana with all rights reserved. Figure 3.2.b XXX (Paolo di Uccello's The Miracle of the Desecrated Host, Scene 1) is reproduced here by kind permission of the Ministero per i Beni e le Attività Culturali della Repubblica Italiana/ Galleria Nazionale delle Marche, Urbino, Italy.

NOTES

1. Dell'Abacco's solution is as follows: "Do in this way: if 3 is worth 5, how much is 5 worth? Multiply 5 by 5, which is 25, and divide by 3, you will have 8 1/3. Now you may say: for each 5 of those (paces) of the dog, you have 8 1/3 (paces) of the fox; so the dog approaches the fox 3 1/3 (paces). In how many paces will he (the dog) reach her (the fox) by (covering) 40 paces? Then say: if 5 is worth 3 1/3 for 40, how many will I have? Multiply 5 by 40, which is 200, and divide by 3 1/3. Bring (i.e., reduce) to thirds, thus multiply 3 by 200, which makes 600, and divide by 3 (and) 1/3 , that is 10/3 and then divide 600 in 10, it gives 60. And the dog will do 60 paces before it reaches the fox. And it is done. And the proof is that in 60 paces the fox goes 60, and the dog in 60 paces is worth 100 [i.e., 60 steps of the dog are worth 100 steps of the fox], because three of his (dog's paces) are worth 5 (of the fox); therefore 60 paces (of the dog) are worth a good 100 (of the fox). It is done." (Arrighi, 1964, p. 78). I am grateful to Jens Høyrup, Fulvia Furinghetti and Giorgio Santi for translating dell'Abbaco's problem into English and for their precious help in the analysis of the solution.

2. Since the pioneering work of Clement (1989) and Disessa, Hammer, Sherrin, and Kolpakowski (1991), informed by cognitive science and constructivism, recent work includes Arzarello and Robutti (2004), Arzarello (2006), Nemirovsky (2003), and Roth and Lee (2004), inspired by embodied psychology.

3. Similar problems and solutions can also be found in many other manuscripts, for example, in the thirteenth century Fibonacci's Liber Abacci (Sigler, 2002).

4. The fourteenth century mathematicians at Merton College in Oxford did not deal with two moving bodies, but rather with theoretical investigations of uniform and non-uniform speed (see Clagett, 1959).

REFERENCES

Alcuin. (2005). *Problems to sharpen the young* (8th century. R. Franci, Ed.). Pisa: Edizioni ETS.

Aristotle. (350 B.C.E). *Physics* (R. P. Hardie & R. K. Gaye, Trans.). Raleigh NC: Alex Catalogue. Retrieved October 6, 2005, from http://classics.mit.edu //Aristotle/physics.html

Arrighi, G. (Ed.). (1964). *Paolo dell'Abbaco: trattato d'aritmetica* [Paolo dell'Abbaco: Treatise of arithmetic]. Pisa: Domus Galileana.

Arzarello, F., & Robutti, O. (2004). Approaching functions through motion experiments. *Educational Studies in Mathematics, 57*(3), CD-Rom, chapter 1.

Bakhtin, M. M. (1981). *The dialogical imagination*. Austin: University of Texas Press.

Bakhtin, M. M. (1986). *Speech genres and other late essays*. Austin: University of Texas Press.

Boero, P., Pedemonte, B., & Robotti, E. (1997). Approaching theoretical knowledge through voices and echoes: a Vygotskian perspective. In *Proceedings of the 21st International Conference for the Psychology of Mathematics Education* (Vol. 2, pp. 81–88). Lahti, Finland: PME.

Brown, M. P. (1990). *A guide to Western historical scripts from Antiquity to 1600.* Toronto: University of Toronto Press.

Clagett, M. (1959). *The science of mechanics in the Middle Ages.* Madison: University of Wisconsin Press.

Clement (1989). The concept of variation and misconceptions in Cartesian graphing. *Focus on Learning Problems in Mathematics, 11,* 77–87.

Disessa, A., Hammer, D., Sherrin, B., & Kolpakowski, T. (1991). Inventing graphing: meta-representational expertise in children. *Journal of Mathematical Behavior, 10,* 117–160.

Galileo, G. (1638), *Two new sciences* (Translated by S. Drake, 1989). Toronto: Wall & Thomson.

Gallese, V., & Lakoff, G. (2005). The brain's concepts: The role of the sensory-motor system in conceptual knowledge. *Cognitive Neuropsychology, 22,* 455–479.

Grize, J.-B. (2005). Time of soft ideas. In A.-N. Perret-Clermont (Ed.), Thinking Time (pp. 68–72). Goettingen: Hogrefe.

Koyré, A. (1966). *Études d'histoire de la pensée scientifique.* Paris: Presses universitaires de France.

Miranda, I., Radford, L., Guzmán, J. (2007). Interpretación de gráficas cartesianas sobre el movimiento desde el punto de vista de la teoría de la objetivación. *Educación Matemática, 19*(3), 1-26.

Nemirovsky, R. (2003). Perceptuo-motor activity and imagination in mathematics learning. In N. Pateman, B. Dougherty & J. Zilliox (Eds.), *Proceedings 27th Conference of the International Group for the Psychology of Mathematics Education* (Vol. 1, pp. 103–135). Honolulu, USA: PME.

Radford, L. (2003). Gestures, speech and the sprouting of signs. *Mathematical Thinking and Learning, 5,* 37-70.

Radford, L. (2006). The cultural-epistomological conditions of the emergence of algebraic symbolism. In F. Furinghetti, S. Kaijser & C. Tzanakis (Eds.), *Proceedings of the 2004 Conference of the International Study Group on the Relations between the History and Pedagogy of Mathematics & ESU 4 - Revised edition* (pp. 509–524). Uppsala, Sweden.

Roth, W. -M. (2003). *Toward an anthropology of graphing : Semiotic and activity-theoretic perspectives.* Dordrecht, The Netherlands: Kluwer Academic Publishers.

Roth, W.-M., & Lee, Y. J. (2004). Interpreting unfamiliar graphs: A generative activity theoretical model. *Educational Studies in Mathematics, 57,* 265–290.

Sigler, L. E. (2002) (Ed.). *Fibonacci's Liber Abaci.* New York: Springer.

CHAPTER 4

WHAT MAKES A CUBE A CUBE?

Contingency in Abstract, Concrete, Cultural and Bodily Mathematical Knowings

Jean-François Maheux,
Jennifer S. Thom, and Wolff-Michael Roth

When presented with the image in Figure 4.1, a person will more often than not immediately see a cube. This happens despite the fact that the drawing *is not* a cube at all: As the French painter Magritte coined many years ago, the representation of the object is not the object itself. So how is this experience of a cube in a drawing possible?

From the viewpoint of phenomenology of perception, to recognize the figure as a cube, one must experience it bodily (looking at it, eying the image from every angle), and

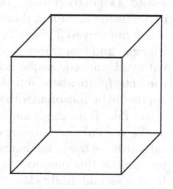

Figure 4.1.
This is not a cube.

Mathematical Representation at the Interface of Body and Culture, pp. 71–93
Copyright © 2009 by Information Age Publishing

71

connect the perception to the culturally defined (mathematical) idea of the cube (Merleau-Ponty, 1945). Contemporary phenomenological cognitive science, too, asserts the physical body as the center from which all knowing emerges, while it defines culture as an open network of ways of dealing with the material world and with others, preserved from generation to generation (e.g., Gallese, 2003). In this view, bodily and cultural knowings are mutually constitutive. It is our cultural knowledge that tells us the drawing is a cube, as we recognize in the image a representation typical in Western culture. However, by looking at the image, we not only can recognize a cube, but we also experience three-dimensionality by bodily knowing the relationship between the ink traces on the flat paper and a block on our desk. Indeed, when presented with a similar figure, young children express their puzzlement: "That's not a cube, there's no triangle on a cube!" For the cube to be seen, the physical body needs the work of culture, and vice versa.

When it comes to name the experience in which body and culture are at play in seeing a figure as a cube, another, yet similar, phenomenon also takes place. It is common for us to consider that the cube and its representation are essentially different kinds of entities. In geometry, a cube is an ideal (transcendental) object, whereas its representation bears all the imperfections of its materiality. What we are less familiar to recognize is again the co-emergence between these two dimensions of the experience of a cube. To emerge means to rise out of or up, as something coming out of a blend, e-merge. The ideal mathematical object is never experienced *per se*, but always realized in some tangible instantiations: a drawing, a block, a word, a set of coordinates, or even a mental image. In the same way, a bloc or a drawing are not in themselves related to cubes. The figure displayed above can very well illustrate a certain pattern composed of a square, two triangles and four trapezoids, and the concept of a cube need not be present at all. Indeed, a cube is for us simultaneously abstract (general) and concrete (particular), and these two dimensions need to draw on one another to be consistent with what we recognize as the experience of a cube.

We take from these and similar observations that abstract, concrete, bodily and cultural knowings in mathematics all co-emerge in the practical activity of doing mathematics. Is this actually the case? In this chapter, we explore this question by presenting and examining episodes taken from a second-grade class in which students are to learn geometry. We begin by reviewing the concepts of the abstract and concrete, the body and culture, in relation to mathematical understandings and then locate and discuss the presence of these concepts in the children's mathematical activity. This takes us to understand how forms of knowing that emerge from the children's mathematical activity are, inherently and contingently, all at once concrete, abstract, bodily and cultural in nature.

ON FORMS OF MATHEMATICAL KNOWING

The Double Ascension of the Abstract and the Concrete

We just saw that a cube appears to us as simultaneously abstract and concrete. Recent work in mathematics education which draws on the writings of Lev Vygotsky has come to address this situation from a dialectical perspective in which concrete and abstract understandings co-emerge from one another and co-evolve as double ascensions (Roth & Hwang, 2006). Moreover, such ascensions are neither directional nor hierarchical. Rather, understanding evolves simultaneously from the abstract to the concrete and from the concrete to the abstract. We represent this by the symbol " | " when we point to the *abstract|concrete* dialectic.

In its etymology as well as in its usage, the term "abstraction" is both a process (the action of drawing out from a situation) and an object (the product, the concept). In the ascension from the concrete to the abstract, a complicated phenomenon is transformed into a simple thinkable unit by constructing relationships amongst the situational objects. For example, to recognize a cube in Figure 4.1, one must focus on a specific part of his or her rich and complex environment. Taking apart the concrete aspect of the figure (black ink and white paper), we recognize the straight segments and a particular organization that the abstract idea of the cube encapsulates. Because it centers thinking and simultaneously gives meaning to the complex phenomena of perceiving the image, the abstract and the concrete dimensions indeed co-emerge in the environment and in the mind.

Abstractions can also be thought as properties of representations that are interpreted apart from their referents as the lived experience is brought about in consciousness. The reader can now think about cubes without necessarily referring to the specific figures from which the abstraction was generated. But in the ascension from the abstract to the concrete, abstractions are also unfamiliar or complex ideas that are situatively concretized in the practical activity. The concept of cube is also rich and complex as it connects with an inexhaustible number of other concepts. The cube can be defined from a variety of perspectives that differ according to their foci (geometry, algebra, arithmetic, calculus, topology), and each time in more than one way (in terms of faces, of edges, of vertices, as a region in space, or as a polyhedron presenting certain properties). The idea of a cube is general and it reveals the potential to embrace any of different experiences and what the reader sees as a cube when turning to the figure is therefore a form of reduction of specification (Roth & Thom, 2009a). Abstract ideas thus become concrete, particular, and arguably complex. The concept sheds light on what is experienced, as context-relevant aspects are noticed. Moreover, in one of its classical geometric representation (Figure 4.1), the idea of cube makes room for the interpretation of the two-dimensional figure to be perceived three-dimensionally. This occurs in the moment-to-moment transactions[1] with, in this case, an object in the material world. Furthermore mathematical understandings (of a figure, for example) are realized by this movement in the conceptual space, when the *abstract*|*concrete* dialectic gives meaning to the practical activity of looking at the figure.

Coming across drawings similar to Figure 4.1 in various contexts also lays a path for the development of conceptual thinking. As part of that movement, the ascension from the abstract to the concrete plays out when what was first unfamiliar, unspecified, entangled in the generality of an undeveloped (abstract) idea becomes more concrete. Through a variety of experiences, abstract understandings turn to be more context-related, precise, rich in details, and so forth. Simultaneously, growth in understanding is also characterized by the ascension from the concrete to the abstract, when a person develops fluency in connecting ideas to one another and drawing out of contextual events (including sensorial experiences) what appears to be relevant to the situation. As the thoughts gain in independence and generality, we reconstruct the developmental trajectory from concrete to abstract forms of thinking. For instance, the idea of a cube and the figure probably "work well" together for the reader, but it is not the case for young children (whose early contacts with cubes are more connected to "square blocks," "ice cubes," and other more or less cube-like objects rather than orthographic projections). The process through which each particular experience contributes in the development of the general idea that embraces all these representation of cubes is a movement from the concrete to the abstract. Concurrently, we can illustrate the movement from the abstract to the concrete by the children's common use of the word "cube" to talk about oblong rectangular prisms: they only later employ it more precisely, taking in account the cube's property of having six square faces. And only a reader with a rich background in Western mathematics can (a) appreciate in detail the interplay of structure within Figure 4.1 and (b) the idea of a cube as an ideal object having identical edges and square faces, which is not the case in the drawing (four of the faces are parallelograms and we have two different sizes for the edges).

BODY, CULTURE AND THE NATURE OF MATHEMATICS

Researchers concerned with mathematics education have been successful in establishing theoretical perspectives that substitute traditional views of mathematics as transcendental, external, and objective. These other points of view make salient the fundamentally lived dimension of mathematics itself, and its conceptualization on the part of learners. The idea of a cube always comes to being in some context (like in reading this chapter or in a classroom conversation) because it is fundamentally something that we humans do. We learn to recognize the line drawing as a cube by observing the picture and discussing it, by drawing it ourselves and connecting all these experiences to others in which the idea of the cube is brought about. We also connect these experiences to a variety of individual or collective endeavors, inevitably situated in specific, singular, sociomaterial contexts. When looking at the figure in this context, the reader is again situated in an infinite web of relationships with people and material things that specify why and how the figure is identified as a cube (Roth & Thom, 2009b). In the context of a discussion of patterns, for example, the same image would be more likely read as a composition of two-dimensional shapes because of the reader's orientation.

Research on the embodied nature of mathematics shows that mathematics is fundamentally grounded in human experiences, naturally arising out of everyday experiences of the world as a result of normal human cognition (Lakoff & Núñez, 2000). In this sense, doing mathematics calls upon the body as to the way we deal with the material world, in the continuous flow of making sense of the situations in which we find ourselves. A phenomenology of perception of geometrical figures (Roth, 2009) suggests that when eyes move about the figure without even "thinking" about it (i.e., in an unconscious manner), the movement leads to the recognition the straight lines, the corners, and the organization of the line segments in which some are parallel whereas others are perpendicular to one another. Visualizing the three-dimensional cube also includes an embodied knowing of the spatialization of the cube's faces.[2] While comprehending all of this, we knowingly disregard the fact that some of the lines cross other lines in the drawing and in doing so, we "see" the faces as if the cube was a three-dimensional object. This knowing-how to view and interpret the image constitutes the embodied knowing in the sense that the eyes know what to do, how to interpret the figure for us to see the three-dimensional object (and not only to think about the idea of a cube). It is crucial to note that these embodied knowings are also essentially cultural. For example, studies with Uzbekistani peasants (Vygotsky & Luria, 1993) show that some people who live in very poor villages and never go to school do not recognize shapes in the same way that schooled individuals do, having grown up and living in Western societies.

Beside that, the figure is still open for interpretation and needs the work of the body in the active perception of the ink traces on the paper. Focusing the eyes on the lower or the upper square shape can bring about two different cubes: one protruding to the bottom left, and one to the upper right. These kinds of optical experiences (studied, among others, by Vygotsky and his colleagues) are used in a phenomenological perspective to show how the body (movement of the eyes) plays out in what emerges into consciousness. Furthermore, identifying line segments and shapes, spatially coordinating them in either 2-D or 3-D and gathering that flow of visual information are examples of how everyday experiences of the world specify one's mathematical experience (seeing or interpreting) of the figure.

In addition, because doing mathematics is not just any kind of sense-making activity but one that has a specific history (that students continue nowadays), culture plays a central role in how what a person does is recognized as (or turned into) a mathematical activity. Culture consists both of established observable practices and artifacts/tools and of an open-ended, unfinished set of possibilities afforded and specified by particular artifacts, like materials objects, tools, language, and so forth. When these possibilities realize themselves in communicating, some of these possibilities have historically come to establish themselves as genres. Communicating (which includes talking, gesturing, and other communicative resources such as body orientation) is both an embodied and a cultural practice. This is so because communicative resources are intelligible bodily productions *for others*, who are presupposed to see, hear, and understand what is meant without thinking and reflecting. Indeed, if a complex genre like mathematics distinguishes some human undertakings from others, it also draws on and is realized in living communication. Consequently, even what seems to be the culturally independent nature of mathematics is actually at the heart of a mathematical culture (van Oers, 2001), a domain in which people like us acknowledge ways of doing things as mathematical (or not). If the idea of a cube potentially has the same (mathematical) consequences for everyone in the world, regardless of their culture in the general sense, it is because they can all draw on and contribute in a mathematical culture.

Cultures exist as they are lived. They are continuously produced and reproduced in practical activities. In the transactions between individual consciousness and the outer social world, the individual and the social re?exively not only complement, but constitute each other: conscious always is consciousness-for-the-other as it is consciousness-for-the-self (Vygotsky, 1987). In this situation, doing mathematics is a culturally enabled way of being in the world. It is historically preserved by introducing students to a cultural way of doing things, and by inducting the students' activities as part of their culture. This explains the generational (or inter-generational) continuity, as well as the dynamic, evolving nature of cultures. We recognize a cube in Figure 4.1 by relying on a range of previous experiences of similar figures: they were seen in textbooks, on working sheets or exams, on chalkboards, and so on. Most of them were part of societally organized activities in which a number of elements contributed in associating these situations with mathematics. These activities were also going on as such because there were people to bring them into being in a particular way. On another level, looking at the figure within this book also illustrates how culture evolves in our continuous conversations: We offer here a new usage of the image in a discussion that contributes in producing, reproducing, and transforming mathematics as a subject matter. The same image also has a history in the development of projective geometry, including its recent utilization in computer vision. One of the most important consequences of this is that the culture of mathematics is not limited to what mathematicians draw on and do, but includes all mathematically related activities, like teaching and learning. All those who in one way or another contribute to the reproduction and transformation of the mathematical genre, using mathematical ideas to achieve their tasks (should it be a classroom session, the design of a computer game or the reading of this chapter) contribute to and are part of that mathematical culture. It is only in and through this continual production of mathematics that the field also reproduces itself.

To summarize, mathematics exists as a cultural domain only as far as individuals, bodily engaged in mathematical activities, transact with the material world and with others: from which abstract and concrete (mathematical) understanding co-emerge. In the following sections, we develop and illustrate these ideas by looking at an excerpt from a lesson in a second-grade classroom. The lessons were designed to allow students to learn geometry. Turning to the teacher-student transactions, we show how the concrete and abstract happen to be part in the consideration of what makes a cube a cube. To that end, we first clarify the concept of abstract, concrete, bodily and cultural knowings in this particular context. Second, we examine the students' transactions with cube-like blocks and how they realized these transactions with and for each other.

WHAT MAKES A CUBE A CUBE?

A Classroom Episode

The following episode takes place in the second lesson of a series of 15 classroom sessions. That day, one of the teacher-researchers (Jennifer Thom) introduced the following task: Each group of three students received a block and then was sent on a search in the classroom to "try and find as many things as [they] can that are the same." Next, they were to "share what [they] found, and explain how [the objects] are all the same thing." Eugene, Bert, and Chi-Chi received a wooden cube-like block. As they searched the classroom, they collected various objects: A red cardboard box, some wooden blocks of various sizes, a big yellow plastic cube and three bundles of post-it-notes (not cube-like). In the second part of the lesson, once all the students were back from the search, Jennifer used a chart where mathematical names and pictures of solids were printed to ask each group to "figure out [the] object's name." A few minutes later, she asked the members of Eugene's group to place their objects on a piece of paper. Going through the following episode, we exemplify how bodily and cultural knowings are inherent in the double ascension of the students' concrete and abstract understandings as they, and the teacher, discuss what makes a cube a cube.

01   Je: Can you put all your things [a] on the purple mat? Okay and take a look at all these things that they say and…. Bert's group says that these are all?
[a] Eugene, Bert and Chi-Chi start placing their objects on the paper. They carefully place them, move them around and position the bigger ones in a way that makes the whole collection visible on the mat.

02   Eu: [b] Cubes
[b] Picks up a yellow cube, holds it in his hand, turns it around, and places it back. Bert and Chi-Chi are still changing the positions of the objects.

03   Je: These are all …

04   Eu: Cubes

05   Je: Cubes, okay. [c] And everybody I want you to look at theirs and they [d] say they're all cubes and a little bit explain to us what makes a cube [e a cube, so Eugene, do you want to start? e] What's one thing that makes a cube a cube [f].
[c] Jennifer pauses. Chi-Chi stops moving the objects, Bert is still turning around a blue piece of paper.
[d] Bert places the paper on the yellow cube and then stares at the collection.
[e] Eugene turns his gaze to the collection, and then looks at Jennifer. Eugene still looks at her while placing his hand on the top of the big yellow cube, holding it by two sides with his fingers and the thumb. At the same time, Bert had turned his gaze to Jennifer and then quickly back to the collection.[f] Eugene picks up the yellow cube.

06   Eu: [g] All the sides [h] it has
[g] Eugene holds the block and turns it around, placing his right hand flat on some of its sides. Chi-Chi stops manipulating the other objects and turns her gaze to the block.
[h] Bert turns his gaze to the cube Eugene is holding and turning around.

07   Je: Sorry? [i]
[i] The three students turn their gaze to her, and then back to the block Eugene is still holding.

08   Eu: All the sides it has, like all the same, they are all the same [j], each side
[j] Eugene turns the object, touching and presenting different sides.

09   Je: Okay [k] so you're saying all the sides are the same [l], what do you mean by same?
            [k] Eugene shortly turns his gaze to the object. Chi-Chi turns her gaze to Jennifer.
            [l] Chi-Chi turns her gaze to the collection of objects on the mat. Eugene is still rotating the object in his hands.

10   Eu: Well [m] each one is the [n] same size and each one is the same the same [o] like square, like, right
            [m] Eugene stops rotating the object, holds it with his hand round it, and looks in the teacher's direction.
            [n] Eugene touches one side/face with his finger, turns the object and touches another side/face in the same way
            [o] Eugene slides his finger all around the edges of different faces.

11   Je: Okay so Eugene says they are all the same size and they are all square [p] Oscar?
            [p] Oscar raises his hand. Eugene returns the cube to the collection.

12   Os: Hum [q] well the squares all have four corners and [r] this square has four corners [s] and all the squares here have four corners and so … and all they have four corners.
            [q] Takes the yellow block from Eugene and look at it, the points to the collection and turn his glance back to the block, touching different vertices with the tip of his fingers.
            [r] Puts down the yellow block in front of him, picks up a small one and touch an edge.
            [s] Return the small block and picks up the yellow one again, using it to points to the whole collection various blocks before he deposits it back onto the paper mat.

Concrete and Abstract From Everyday Knowings in the Mathematics Classroom

We are positioned in our everyday lives within an established system of experiencing the material world (e.g. seeing, touching) that links what emerges to the sense with a complex web of previous experiences (visual or tactile) and with a set of culturally possible ones (examining a drawing, manipulating a block). When brought about in the practical activity of manipulating objects and engaging in a classroom conversation, the idea of the cube simultaneously constitutes an abstract and a concrete form of knowing. When the students move around the blocks, touch and refer to them, they use concrete instantiations of a cube, materialized "rough approximations" of the ideal concept. At the same time, when the students use the objects, name them "cubes" (turn 02) and discus what makes a cube a cube, they enter in the domain of abstraction. In this episode, Eugene does not attend to certain aspects of the blocks (color, material, size, functions in another context) but he does explicate other properties that appear to be relevant to the situation (e.g., the fact that they have square faces, turn 10).

We also see the movement in the *concrete|abstract* dialectic in this brief episode through the transactions between Eugene and Jennifer. Eugene translates the appreciation of consistency in the block (concrete) in terms of sameness ("they are all the same, each side"), an abstract, general idea. In turn 09, as Jennifer asked him to explain what he means by "the same," Eugene moves forward, and the abstract idea becomes more precise, concrete: "each one is the same size and each one is ... square." We learn what is a cube as an abstract concept and in its concrete instantiations by engaging in transactions with material objects and by articulating ideas about them. Growth of understanding is therefore something that occurs in the realization of our ongoing activities (naming objects, moving them on a mat, explaining why they can be called cubes). It is from everyday embodied knowing that abstract mathematical concepts come to life, while these concepts are always connected to these everyday experiences within a specific cultural domain of interaction, a genre, like geometry.

The role of body and culture in the practical activity can also be observed in the episode. In geometry, a cube is an ideal mathematical entity, a regular solid, a region of space formed by six identical square flat faces orthogonally joined along their edges. But to know "cubes" is much more than being able to produce a definition. It is to demonstrate a competency to use words, interpret ink traces on a piece of paper or blocks, to act according to the mathematical properties of a cube as to a given activity, and so on. There is also a form of embodied knowing about cubes that comes with the experiences of the blocks. To know an object in its cube-like dimension is to know how it affects the senses, what these impressions reveal about the structure of the object and how they are consistent with the idea of a cube. In other words, it is to know how to hold a cube-like block, how it feels and how to manipulate it. Accordingly, we recognize forms of bodily mathematical knowing when students bring about salient aspects of mathematical thinking as they use their body to carry out a task.

In the episode, the concrete experience of the block perceptually guides Eugene to develop the idea of a cube. Along with his first manipulations, Eugene indicates that a cube is an object that has sides (turn 02). In return, it seems that his conceptual understanding guides his perception, focusing his attention to the faces themselves, so that he can see and feel the conservation of distance between the edges in all the faces and conclude they are all the same size, and all square. Physical manipulation is the way in which Eugene explores the object, and this exploration brings forth his articulation of the idea of what makes a cube a cube. It is the conservation of the haptic and the visual impressions of the object (in turning it around) that can give rise to the observation of sameness. To rephrase, experiencing different faces in different orientations reveals that they all look or feel the same, in whichever orientation they are. In the practical activity of dealing with the block, Eugene already realized the concretization of an abstract idea: the block he picks up (gesture [b]) is a member of the "object" category, something tangible that can be held and touched. In the same movement, turning his attention to the block, Eugene experiences a concrete, material instantiation of a cube while talking about it in an abstract way (explaining that what makes a cube a cube is all the faces that it has). The experience of the block then develops in both the concrete and the abstract dimensions, as Eugene feels and articulates more. Indeed, he touches and mentions other aspects relevant to the concept, talking about dimension ("size") and shape ("square") of the faces (turn 10). Perceptually and conceptually guided and guiding his transactions with the block, the abstract becomes more abstract, the concrete gains in concreteness, and not only simultaneously, but literally from one another. It is because Eugene is in a position to feel more in the concrete that he can articulate more in the abstract, and vice versa.

In examining the previous episode, we see that these embodied forms of knowing are related to everyday experiences of the world: It is from very early experiences in life that we develop a sense of space, boundaries, organization, and three-dimensionality of objects. These are not a random collection of experiences, but rather, they are the result of an established system of experiencing the material world. And we need to remember that they are at the same time constitutive parts of establishing that system from generation to generation. The students carefully pick up the objects and move them, limiting their exploration to the visual and haptic dimensions, while they could have tried, for example, to taste them, smell them, throw them away, or even break them apart. Therefore, they demonstrate cultural knowing about how we generally treat certain objects and in certain contexts. In his contribution, Eugene draws on everyday knowing about objects: picking up the yellow cube and turning it around is not problematic (like it would be the case if, for example, the sense of touch was failing him). This form of knowing is culturally oriented: Eugene holds the block in ways that differ from how we would, for example, touch a living being. Further, these embodied forms of knowing (drawing on everyday experience of the world) allow the students to engage in a mathematical genre (a way to deal with objects and communicate understanding) by focusing on the description of the blocks, looking for regularities (like Eugene saying that all the faces are the same) or similarities (like Oscar explaining that all squares have four corners).

In other words, our episode exemplifies how embodied forms of knowing are fundamentally cultural (developing from earliest experiences), while all cultural forms of knowing are realized in the body (manipulating objects, writing, or reading a definition in a text book). Because it is impossible to recall and connect all these experiences (that might or might not be relevant to the activity), a concept is at the same time both abstract (what is possible) and concrete (what is actually brought about). These inseparable ways of knowing are not the result of some mind game, or of uncommon phenomenological experiences. It is something students, as all people, do at every moment in the practical activity of doing school and as an integral part of what learning mathematics is.

Concrete and Abstract in Transactions With the Material World and With Others

The episode exemplifies how, in relation with the material world, the cube as a mathematical object is experienced bodily and that these experiences are culturally oriented. Articulating *concrete|abstract* observations realize culture in and through the body. But to go beyond these general assertions, a close observation of what is going on in the episode is necessary. To fully appreciate the complexity of students' mathematics, in the following sections, we (a) exemplify how sensual, cultural, concrete and abstract experiences are translated in each other by the students and (b) conceptualize how the student's transactions with the material world are realized in and for communication with others. This helps us understand how bodily, cultural, abstract|concrete forms of knowing in mathematics are all fundamentally contingent on one another. We illustrate these relations by (c) examining the role of cultural artifacts in bringing about mathematical ways of dealing with objects and ideas.

The Translation of Sensual, Cultural, Concrete|Abstract Experiences

When students manipulate objects while discussing what makes a cube a cube, a translation takes place between bodily and cultural forms of knowing. To illustrate this, there are limits to what can be said about the visual experiences of the objects in natural settings: movements of the eyes, changes in focal point, and so forth are not accessible through the data we collected. However, to get a sense of how these translations are at play, we can turn to the students' gestures and examine how they manipulate the objects. Looking closely at the students' transactions with the objects allows us to develop the concept of genre as key to articulate the cultural dimension of the students' practical activities. In the analysis of genre in communication (i.e., forms of communicating), the utterance, which includes the response received on the part of the producer, was chosen as the minimal analytic unit (Bakhtin, 1986). Utterances are comprised of both speech and gestures. They are the rough, lived material from which and through which a genre realizes and concretizes a particular way of dealing with the world and with others. Students' utterances are what brings culture into being, and at the same time culture is what makes any intelligible utterance possible, that is, something a student *intelligibly* does. Paraphrasing Bakhtin we might say that it is through concrete utterances that culture enters life and that life enters culture.

In our excerpt, Eugene's manipulation of the cube is rich and complex (Figure 4.2), unfolding with his verbal articulation of the properties of a cube. From a phenomenological perspective, the block in itself cannot be known, as one can only identify the effect that it has on him or her. As he reaches for the yellow block (gesture [b] and [e], Figure 4.2), Eugene can experience volume and can feel resistance. In gesture [g] and [m] (Figure 4.2), he uses his two hands to feel the surface of four of the six faces of the block which allows him to experience their flatness and their boundaries. Moreover, holding a cube requires the complex coordination of the arms, the hands, the fingers, and the continuous interpretation of sensory information. In picking up the block, Eugene shows an embodied form of knowing about the cube as a solid: a region in space bounded by faces.

Using a block to talk about cubes is also intertwined in the cultural dimension of mathematical forms of knowing. We know that there is a need to select among what aspects of the object are felt and what is to be drawn out of that sensual experience about what is relevant in the cultural domain in which these observations are made. It therefore appears to us that Eugene translates his sensorial experiences within the cultural domain. The weight of the block, its color, and the particular volume it occupies are examples of the object (and therefore objective) characteristics that he does not draw attention to. These translations reveal a form of participation in a cultural-mathematical way of considering the material world. They realize a way of talking about (and manipulating) objects, to make sense of experiences that contextualize a cultural possibility directly related to what we define as doing geometry.[3]

The translation of sensorial and cultural knowing goes both ways. A cube, explains Eugene, is an object whose faces are all the same size and all square, which is something he can articulate without entirely knowing

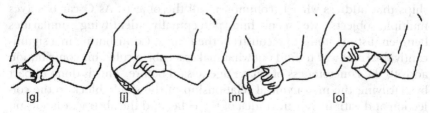

[g]   [j]   [m]   [o]

Figure 4.2.   Four ways of touching the yellow block in gesture [g], [j], [m], and [o].

"all" that makes a cube a cube. From a rich and messy background of bodily and cultural knowing about objects and linguistic terminology— the words "cube," "sides," and "square" that Eugene uses without the need, at this point, to define them—surfaces the idea of a cube. Whereas what the body realizes is a cultural (mathematical) orientation, the genre itself takes part in orienting the sensorial experiences. For instance, cognitive phenomenological studies show that the fact that the light from an object falls onto the retina cannot guarantee that the person consciously sees something (Roth, 2005). Not only does culture guide bodily actions in the material world (and in the immaterial world of ideas), but also the body realizes culture in turn.

Picking up one block amongst others to talk about the whole creates a network of representations in which the abstract and the concrete are at play in a specific way. The particular block represents the whole collection, and both are concrete entities in relation to the abstract idea of the cube. Manipulating and talking about the block, Eugene allows us to witness his search for "internal" relationships. From this perspective, saying what makes a cube a cube is to look for perceptual differences from a whole (the concrete bloc) to reconstruct the unity with other objects and thus access its abstract dimension in the mind and in communication. The idea of the cube, reified in Eugene's talk (turn 02 and 04, as he answer "Cubes" to Jennifer's question) to designate the collected objects, is concretized as Eugene explains what makes all these objects cubes. It is as if Eugene is translating his concrete experiences of the block in a domain of abstraction (and vice versa). An abstract understanding of cubes is demonstrated by concretely selecting a block from the collected ones to talk about cubes in general. Indeed, one assumes in such a case that what is pointed to and felt (as essential elements of what makes a cube a cube) can be experienced with any from one collection. The particular cube Eugene holds becomes the general cube in relation to the collection while at the same time it addresses the abstract idea of a cube. A similar analysis can be made of Oscars' utterance in the last turn of the episode. Oscar manipulates several blocks as he explains that "all the squares ... have four corners." In this case, looking at different blocks to answer Jennifer's question is about identifying a set of "external" relationships that address what is common to all the objects. As Oscar observes multiple objects, we seem him perceptually identifying similarities between distinct items to reconstruct their unity. Contributing in a significantly different way to find out what makes a cube a cube, his contribution adds to the concreteness of what Eugene said before. This is done in part by clarifying the metonymical relationship of the single block to the collection and with the abstract idea of the cube, and the abstract dimension of what is brought about which also deepens the process.

Communication With and for Oneself and Others

In the last section, our analysis of the children's mathematical activity draws on what can be found in both their speech and gestures. We look at the words they used and the way they manipulated the objects at hand. These are constitutive elements of communication engaging the whole body. For example, when Eugene or Oscar communicated ideas about what makes a cube a cube, they produced sounds with their vocal cords and the movement of their thorax, tongue, jaw and lips, and they also used their hands to touch and point to the blocks. However, the effective meaning these resources bring about in conversation cannot be detached from the immediate context in which they are produced. What might look like a similar utterance ("I am doing well") can have a very different meaning according to the situation (such as talking to a neighbor or to a doctor). For this reason, an utterance should not be considered in isolation, but rather in relation to what comes before and after it, to understand what it answers and how it is responded to (Bakhtin, 1986).[4] As a sense-making activity, communication also presupposes intelligibility, that is, the engagement of a listener (that can be oneself or another) for whom the utterance is produced, and whose response completes it. In addition, the cultural aspect of communication is obvious in both the utilization of words and in "body language."[5]

Because each contribution is woven into the unfolding classroom discussion, Eugene's and Oscar's communication were created together with the teacher and the other students. And it is precisely in this collective dimension that the meaning of students' utterances reveals itself. As part of the classroom lesson in geometry, for the collective purpose of understanding what makes a cube a cube, each utterance is a response to what had previously been said or done. This is exemplified in the events that followed.

As the objects are moved to the mat, Eugene turns his gaze to the blocks, and then to the teacher as she starts talking. Although Eugene, in response to Jennifer's question about what the objects are, seems to solely address his answer to her ("cubes," he says, looking in her direction), the teacher's utterances show that the conversation is not intended to be a one-on-one dialogue, but rather to engage the members of the entire classroom ("*everybody* I want you to look"). What we see happening here in communication is a re-orientation of the students' engagement from the individual experience of the blocks to a collective endeavor of sense making about them. The orientation of the students towards the collective endeavor is proven successful by the change of the students' gaze during the episode. In the beginning of the excerpt, Chi-Chi and Bert are looking at the collection until the moment when Eugene actually picks up the big yellow block and starts talking (turn 02). At this moment, they turn their gaze to the object Eugene is holding (Figure 4.3). As the conversation continues, the students' gaze then moves repeatedly from the object Eugene is holding to the blocks on the floor and back to the teacher. The change in the orientation suggests that the students are now attending to the classroom discussion. In its collective dimension, the task reveals itself as a cultural one, something the teacher and the students do together by taking part in the classroom event they create. Moreover, it is essentially through their body orientation that they realize this, bodily expressing their attention to the discussion over what makes a cube a cube. We therefore see that Eugene's communication about the cube is indeed realized for and with the others. Whereas none of the students appears to participate otherwise at this time, the fact that they do not interrupt Eugene also counts as contributing to the conversation. Moreover, what follows in the classroom discussion confirms that the students actually were, at this moment, attending to what Eugene and Jennifer were communicating. And just like Oscar mentioning that all squares have four corners, other students will soon contribute and *add* something to what had been said.

As the conversation unfolds, the students and teacher's orientations confirm Eugene's contribution as a valid one, progressively entering the speech genre of accountably doing geometry. As a way of transacting with the material world and with others, the genre is not simply imposed on the students, but slowly emerges as such from the transactions between the students, the teacher, and the material setting. For example, a few minutes before our episode, Oscar and the two other

Figure 4.3.   Some students' orientation while Eugene explains that what makes a cube a cube is "all the sides that it has."

students of his team were also asked to explain why their objects were all cones. Oscar answered the question by telling the story of his search in the classroom: "Because these ones are the same, and then I found this one, and then I found this one and they all [look] like a cone." In contrast, Eugene spontaneously addresses the question using the cube concept. Through each other's response (including the teacher's validation), the genre defines itself within the classroom in a way that is consistent with the broader cultural practice of geometry. We recognize this in the manner that sensual interpretations are translated in verbal or bodily form and subsequently are validated in communication. In turn 05, the teacher repeated the word Eugene used to name the objects ("Cubes, Okay") and then used it four times to ask a new question about the collection ("what makes a cube a cube"). Similarly, it is the conversation that takes place in the collective that confirms Eugene's interpretation, suggesting that his sensual impressions of the cube and their translation in language ("all the sides that it has," "all the same size," "all … square") are culturally valid ones.

The same can be said about gestures that will shortly become a favored way of communicating about solids in the classroom, like Eugene's flat-handed gesture to indicate a face. It is from these kinds of sensual experiences that abstract ideas are developed as a contextual way to make sense of the world—of what is felt, seen, and heard. Simultaneously, in the moment-to-moment transactions with others, embodied knowings become legitimated as specific ways of being in the world. They conceptualize in their similarities what is experienced in different contexts, such as a classroom search for "things that are the same" or a classroom discussion about what makes a cube a cube.

Because what happens in our classroom episode enters the realm of communication, an understanding of the concrete and abstract co-emergence needs to be closely examined from the perspective of how meaning arises from the interpersonal transactions at play. We briefly re-examine the following three turns in the conversation:

08 Eu: All the sides it has, like all the same, they are all the same, each side
09 Je: Okay so you're saying all the sides are the same, what do you mean by same?
10 Eu: Well each one is the same size and each one is the same, the same like square, like, right
11 Je: Okay so Eugene says they are all the same size and they are all square. Oscar?

Our first observation is that we know what Eugene is talking about because his utterance, in turn 08, is offered as an answer to Jennifer's question ("What's one thing that makes a cube a cube"). Similarly, if we understand that Jennifer positively received Eugene's answer (as answering her previous question) it is because we have access to how, in turn 09, she responds to him ("Okay so you're saying all the sides are the same"). Second, this brief exchange between Eugene and Jennifer confirms *within the situation* the ascension from the abstract to the concrete. We interpret what Eugene's utterances potentially mean *for us as researchers* (in which case we detach what we call the student's mathematical activity from the activity itself, as lived that day, in that specific classroom), but also we can see that Jennifer and Eugene experienced it the same way, as the general idea of the cube becomes concrete in the specification of some of its characteristics.

Third, it is again in a simultaneous movement that the concrete, complex experience reaches the abstract realm of thinkable units. The undifferentiated perception of sameness of the block is addressed through the abstract concepts of size and shape ("square"). These concepts emerge in conversation while Eugene answers what he understands as Jennifer's initial question and while Jennifer (in turn 09) responds to Eugene's utterance by stressing one particular aspect of his talk ("what do you mean by same"). Eugene then produces an explanation in which, through Jennifer's reaction, we recognize an observation ("each one is the same size and each one is ... square") of an observation ("they are all the same, each side"). It is because she answers the way she does ("Okay so Eugene says they are all the same size and they are all square") that Eugene's contribution takes part in the conversation as an ascension from the concrete to the abstract in mathematical thinking not only for us, but in the here and now of the children's engagement. This is why we can argue that concrete, abstract, bodily and cultural mathematical knowings are inherently and contingently at play in the children's mathematical activity as *they* experience it, and not only as *we* analytically conceptualize it.

The Role of Cultural Artifacts

When we pay close attention to what is meant by bodily, cultural, abstract and concrete forms of knowing in mathematics, and look at how they are brought forth in what a teacher and her students are doing, their co-implication become increasingly evident. The abstract is always culturally and bodily dependent, just as the concrete is. We can even more clearly illustrate the contingent nature of these knowings by examining the role of cultural artifacts.

When students are doing mathematics, the material objects at hand are of great importance because they contribute to the orientation of the activity. Cultural artifacts are made to preserve and bring about some ways of understanding the world, even though they do not constitute knowledge in itself. However, artifacts do present cultural affordances. An object like the plastic block Eugene is holding is not just any object. As a concrete instantiation of the concept of a cube, its faces were crafted to be flat and smooth, its edges straight, sharp and all the same size, making as salient as possible the properties attributed to the abstract concept. On a larger scale, culture includes the means by which these blocks are produced. Mathematics teaching and learning, as part of that culture, make it possible for the children to enter in transactions with these objects. From there, the material properties of the blocks are experienced bodily by the students in order to enable the *abstract*|*concrete* dialectic we see in the episode, as body and culture work together in the orientation of the students in the task. In contrast, the foam cone used during the first classroom session, where students were sorting objects, is quite illustrative: The cone had an eroded tip, suggesting to the student the creation of a distinct group in comparison to pointy-top cones. This example also illustrates that cultural affordances only exist as long as they are part of a culturally oriented activity. For example, in a calculus class, the same cones could be used to bring about a discussion of differentiability; and in a topology class, both have two faces and a line (circular base), but only one has a vertex whereas the other does not.

It is important to note that not all artifacts have the same orienting power, even though they are all critical tools ensuring that not everything has to be re-invented from generation to generation. Some cultural groups do not use written language and their knowledge is still preserved through oral language and face-to-face intergenerational transactions around the utilization or the production of cultural objects. Nevertheless inscriptions, like the board with names and pictures the teacher used when she asked the students to figure out their objects' names, are very powerful artifacts. For the learning of mathematics, such a phenomenon is widely taken into account by researchers studying technological environments like dynamic geometry software. Eugene's choice to pick up a big yellow block among all the objects is not trivial but significant. Indeed, because of its size, the block was a good choice to sensually experience what is a cube and to communicate these experiences to others. Whereas this might seem to be an isolated event, it was not in this classroom. In the next few minutes, teacher and students would select this same object four times and use it many times during subsequent lessons. Similarly, the chart with names and pictures of solids also was not used for the last time. The teacher recursively invited the students to use it when words "failed" them, up to a point where the students themselves spontaneously turned to it, using it as a tool to enact memory allowing them to adequately engage in the mathematical/geometry genre.

All of our actions are, in one way or another, carried out through cultural artifacts, and language is probably one of the most observable. Language is something produced, reproduced and transformed by culture.

Language exists independently of us (in the sense that it is a social creation that does not depend on any particular individual) while at the same time it only exists through each one of us as we use it (Leontiev, 2005). Words like "cube" or "square" are part of the lexicon of the mathematical genre of geometry. They have a cultural meaning the students learn in the practical activity of employing them, even though they might not, in the beginning, have a clear idea of what they mean. This also illustrates the movement from the abstract to the concrete through (linguistic) artifacts, and the important role that they play in the conservation of our cultural ways to make sense of the world. Although, language is also experienced bodily as we feel things and name them, using our body to talk (producing sounds) about them with other people (describing what we see, what we feel). Speaking and writing are concrete actions by means of which we concretize abstract entities to make descriptions for ourselves and for others.

Ontogenetically, even our body—our whole sensorimotor system—is a cultural artifact. We learn to use our body in transaction with others, in what "forms the sensory composition of specific images of reality—currently perceived or arising in memory, relating to the future, or even merely imagined" (Leontiev, 2005, p. 14). We learn to exist as human beings, to walk, to use objects, because our body is part of a human social network in which these ways of being are constantly produced and reproduced. In this sense, we are ourselves the most powerful of all cultural artifacts, and this explains why human mediation is so important. In turn 05, for example, Jennifer asks the students to look at the objects and try to figure out what makes a cube a cube. In so doing, she orients the students to the material objects, suggesting that an answer to her question might lay in the observation of these artifacts. Despite the fact that she does not mention how she sees them, Jennifer's invitation to look at the blocks suggests that she has a certain way to sensually and conceptually perceive them. Students respond to the invitation, turning their gaze towards the collection (see Figure 4.3). By looking at the objects, students notice relevant aspects, and the blocks serve as a focal point for the initiation of the concrete and abstract double ascension. But to look at something is not just to see it. The objects were already visible to all the students (placed inside the circle formed by the students), Jennifer then invited the students to focus their (visual) attention on the blocks in order to examine them. If (part of) what makes a cube a cube is, somehow, visible in the blocks, this needs to be discerned. By looking at the objects, the students are invited to see them differently, forming new images about them in a bodily, sensual, visual experience. Research in cognitive science has long demonstrated that these experiences of the material world shape the sensorimotor system, enabling us to operate with objects in a specific, culturally oriented way. And here, the features of the cultural artifact guide that orientation so that the transactions (perceiving the blocks, examining them further) bring about mathematical ideas, mathematical ways of dealing with objects and ideas, generating the concrete and the abstract dimensions of these experiences.

THE CO-EXISTENCE, CO-EMERGENCE AND CO-EVOLUTION OF ABSTRACT, CONCRETE, BODILY AND CULTURAL KNOWING IN MATHEMATICS

In this chapter we exhibit the double ascension of concrete and abstract and how this double ascension directly links bodily and cultural forms of knowing. Recognizing and inquiring into the embodied and cultural theoretical and focal concerns, the practical activity shows us how these are nested and seamlessly connected phenomena. Whereas distinct—and therefore irreducible to one another—body and culture are actually dependent, inseparable, and emerging from one another in the transactions with the cultural-material world.

Examining the teacher's and students' practical activity in a classroom, concrete and abstract knowings reveal themselves as forms of bodily and cultural knowings. When Eugene uses a singular object to give access to the cultural domain in which a cube is defined, his manipulations and talk contribute bodily to the cultural genre that defines geometry as a discipline. In the same movement, geometry as a genre provides Eugene an entry point, a way to transact with the block. These transactions—in the relationship they establish with the other blocks and with the idea of a cube—illustrate the *abstract|concrete* dialectic. From the concrete to the abstract, the singular block brings about the general idea of the cube, whereas from the abstract to the concrete, the idea of the cube is what all these objects are leading him (and us) to consider from that specific instantiation about what it is that makes a cube a cube. Briefly, then, teachers and students use their bodies for contributing to and making use of the math cultural domain. They do this in communication and when dealing with cultural artifacts resulting in abstract and concrete mathematical knowings.

If a person looks at Figure 4.1 and sees a cube, it is because bodily knowings work together with cultural knowings in the social context that reading this paper brought about. In this re-cognizing action, the abstract and the concrete co-emerge, and each cognizing instance is a step forward on the path of understanding in which abstract, concrete, bodily and cultural forms of knowing develop. As something we can observe in the everyday work of students in school, (mathematical) meaning (from the

Latin *medianus*, "middle") results from the co-existence, co-emergence, and co-evolution of these forms of knowing in the practical engagement of doing mathematics. Recognizing the contingency of abstract, concrete, bodily and cultural knowing stresses the need to be particularly mindful and attentive to what teachers and students bring into being. And we must acknowledge and value the complexity of these interplays through which students are and become mathematical in the very process of doing mathematics.

ACKNOWLEDGMENTS

This study was made possibly by two research grants (to JT and WMR, respectively) and a doctoral fellowship (to JFM) from the Social Sciences and Humanities Research Council of Canada. We thank the classroom teacher and children for participating; and we are grateful to Mijung Kim and Lilian Pozzer-Ardenghi for their assistance in the data collection.

NOTES

1. A term used in contrast to the concept of *interaction* to implies the irreducibility of a person's experience and of the object of that experience. In that sense, the idea of transaction permits an analytical separation of the two as independent entities that cannot be reduced to one another. What exactly a person sees when he/she looks at an image is the image's response in relation to the person's action of seeing it.
2. We enact such spatialization in everyday life, for example, when we recognize a person we pick up at the airport or train station after having previously seen only a photograph of her/him.
3. In contrast, the meaning of the organization of a pattern, in some African cultures, is developed through the elements of a story rather than by describing it.
4. Only this way can we distinguish between the doctor-patient opening chat and the beginning of the medical consultation.
5. Strictly speaking, the term "body language" is a misnomer, for most body movements (including gestures) have neither syntax nor semantics, the condition for a system denoted by the term "language."

REFERENCES

Bakhtin, M. M. (1986). *Speech genres and other late essays*. Austin: University of Texas Press.

Gallese, V. (2003). The roots of empathy: The shared manifold hypothesis and the neural basis of intersubjectivity. *Psychopathology, 36*, 171–180.

Lakoff, G., & Núñez, R. E. (2000). *Where mathematics comes from*. New York: Basic Books.

Leontiev, A. N. (2005). The structure of consciousness: Sensory fabric, meaning, personal sense. *Journal of Russian and East European Psychology, 43*(5), 14–24.

Merleau-Ponty, M. (1945). *Phénoménologie de la perception*. Paris: Gallimard.

Roth, W.-M. (2005). *Doing qualitative research: Praxis of method*. Rotterdam: Sense.

Roth, W.-M. (2009). Phenomenological and dialectical perspectives on the relation between the general and the particular. In K. Ercikan & W. -M. Roth (Eds.), *Generalization in educational research* (pp. 235–260). New York: Routledge.

Roth, W.-M., & Hwang, S. (2006). Does mathematical learning occur in going from concrete to abstract or in going from abstract to concrete? *Journal of Mathematical Behavior, 25,* 334–344.

Roth, W.-M., & Thom, J. (2009a). Bodily experience and mathematical conceptions: From classical views to a phenomenological reconceptualization. *Educational Studies in Mathematics, 70,* 175–189.

Roth, W.-M., & Thom, J. S. (2009b). The emergence of 3d geometry from children's (teacher-guided) classification tasks. *Journal of the Learning Sciences, 18,* 45–99.

van Oers, B. (2001). Educational forms of initiation in mathematical culture. *Educational Studies in Mathematics, 46,* 59–85.

Vygotsky, L. S. (1987). *Thinking and speech*. New York: Plenum.

Vygotsky, L. S., & Luria, A. (1993). *Studies on the history of behavior, Ape, primitive, and child*. Hillsdale, NJ: Lawrence Erlbaum Associates.

CHAPTER 5

EMBODIED MATHEMATICAL COMMUNICATION AND THE VISIBILITY OF GRAPHICAL FEATURES

Wolff-Michael Roth

But what I perceive in the dawning of an aspect is not a property of the object, but an internal relation between it and other objects. (Wittgenstein, 1958, p. 212^{e})

In the cognitive science and mathematics education literature on graphs and graph-related skills (interpretation, production), it is assumed that certain features (e.g., slope, height) inherently stand out implying that those asked to interpret a graph are assumed to be addressing the sense of these features. The epitome of this approach can be seen from cognitive science research that proposes mental models that include iconic images of the graphs in the human brain (neurons). On the other hand, there are a number of studies among scientists that show how the visibility of graphical features itself is the outcome of a collective process, which elaborates what it is that can be seen objectively. In this chapter, I investigate, using a method grounded in conversation analysis and ethnomethodology, how

Mathematical Representation at the Interface of Body and Culture, pp. 95–121
Copyright © 2009 by Information Age Publishing

graphical features become visible and objective by analyzing the work of pairs of scientists asked to interpret in-field and out-of-field graphs. The pairs have been chosen over individual think-aloud sessions because, in attempting to complete the requested task, the participants make available *for* one another what is required to remain aligned in solving the problem at hand. The study shows that perception includes an essentially passive (rather than intentional) element, where features have to be understood as given to the individual. Any initially individual way of seeing a feature becomes objectively available in and for the community through the embodied communication involving (deictic, iconic) gestures, body orientations, and prosody. For both features of the graph interpretation sessions, the products are better understood as emergent rather than as the outcomes of intentional processes. The purpose of this chapter then is to articulate how graphs and graphical features progressively come to be visible in and through sequentially produced bodily and embodied communication (talk, gestures, body orientations, prosody).

INTRODUCTION

The production and use of graphs is one of the central mathematical practices in the sciences and a pervasive feature of everyday life. It therefore does not surprise that graphs and graphing practices are central components of curriculum and policy documents (NCTM, 2000). Accordingly, all students should be able to create and use inscriptions to organize, record, and communicate mathematical ideas and use inscriptions to model and interpret different material and social phenomena. Graphs are among the quintessential inscriptions scientists use; producing and interpreting graphs therefore also are central to science education (NAS, 1996). Despite this interest and commitment to graphs and graphing, what it takes to interpret a graph is much less clear.

Recent studies of scientists interpreting graphs from their own discipline show that a substantial and surprisingly high number of them experience tremendous problems arriving at interpretations that university professors would accept as correct from their first-year students (Roth & Bowen, 2003). Much of the research on scientists' and technicians' graph interpretations of familiar and unfamiliar graph shows that graphs are sites for articulating what a person already understands, including knowledge of the objects and phenomena denoted, aspects of the data collection and transformation, and familiarity with the instruments and tools required. Graphs that scientists produce on their own, in fact, have metonymic (synecdochical) function, that is, they are the results, and therefore aspects, of a context to which they point back; they are part of the context and used to denote the context, much like a server in the local diner might talk about the "ham sandwich," thereby denoting the person eating a ham sandwich. Thus, when scientists read their own graphs, they begin by providing information about the contexts, tools, interests, research programs, and so on before saying what graphs are about and depict.

Researchers of graphs and graphing generally are interested only in the processes by means of which research participants come to relate graphs and graphical features to their familiar world. However, relating graphs, which are forms of signs, to aspects in the world is only one aspect of interpretation. The second important process seldom attended to by researchers concerns the structuring of the display, a process from which emerge features that serve as signs that are related to objects and processes in the world: this is a perceptual structuring (Roth & Bowen, 2001). This research shows that specific graphical features relevant to scientists' work are not or not necessarily salient to a person unfamiliar with it, even when the graph is from a scientists' domain and even if undergraduate students are supposed to be able to understand the graphs.

In the cognitive sciences, the visibility of the various features that make a graph is taken as unproblematic. Thus, in one modeling effort of graphing, the visual image of the graph to be interpreted exists in the mind of the person (Tabachneck-Schijf, Leonardo, & Simon, 1997); the mind is like a camera—the computational model is called CaMeRa—that mirrors what is materially available in the public space before (in front of) the interpreting persons. The fact that experienced scientists may talk about the slope of a curve rather than its height at different points shows that what is salient and therefore visible is more complicated than the physical image on the retina—given that the eye works optically, we can assume that the rods and cones stand in something like an optical relationship with the material world at which they are directed. The relationship between signs and the world also mediates whether some graphical feature may be stabilized or whether it is treated as a fleeting impression that disappears over time: when a feature (potential sign) does not appear to point to something, it in fact is not a sign and does not become stabilized as such.

Analyses of scientists at work show that when they work with new and often unanticipated phenomena and graphs, the visibility of graphs and graphical features is *accomplished* rather than given; and oftentimes they draw on tools, instruments, and processes to aid in making features visible and relevant (Roth, 2003). These studies also provide us with glimpses into visibility as a social rather than merely physiological process, which is the starting point of this study. Thus, we can talk *about* a feature only if it is materially *there for* and available *to* transaction participants. If participants cannot assume that an aspect is available to other participants, they have to be able to make it visible. That is, a feature is discoursable only if it is, in a sense, objectively present for all; and if it is there and present for all, it is a social object, because it requires processes and competencies that are shared within a culture.

THINKING AND SPEAKING

In this chapter, I take a theoretical perspective on thinking, speaking, and discoursing that is grounded in cultural-historical activity theory and, more specifically, in the work of Lev S. Vygotsky (1986) and Mikhail M. Bakhtin (1986). The former provides us with a theoretical framework of the relationship between thinking and speaking, whereas the latter provides a commensurable framework for communication involving two or more individuals. A combined framework taking into account both speaking or listening, on the one hand, and thinking, on the other, is presented in Figure 5.1.

For Vygotsky, thinking and speaking are two cultural-historical processes that themselves change in and through time. They are part of and therefore mediated by an overarching irreducible unit that he called *word meaning*. Word meaning retains two dimensions: on the one hand, it is utterly singular, because thinking is realized in and through the individual body; and on the other hand, the repeatable moment, the word and its linguistic dimensions, are shared with others in the community. Speaking, too, is an embodied production of sound. Word meaning therefore includes both the particular and the universal, moments that are tied to the individual body and moments that are common with other speakers. Speaking cannot be understood independent of the listener, because it is for the other, whom it takes into account, and to whom language returns. For the listener, a similar relationship between the spoken word and

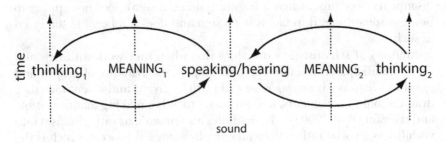

Figure 5.1.   This theoretical framework combines Vygotsky's work on the relation between thinking and speaking and Bakhtin's work on dialogicity.

thinking exists, again subject to word meaning with moments that are universal (structural aspects of language) and particular. For Bakhtin, the speech situation involving speaker and listener cannot be reduced. Although the effect that speaking has on the listener begins with speaking, it becomes available only in and through the responding utterance, which is part of the response as a whole.

Communication involves more than words (sounds); other modes contribute in essential ways in making sense of and understanding what is being communicated. Iconic gestures, because of their similarity with objects and events, can be used to denote entities in communication. Deictic (pointing) gestures also bring physical entities into communication but they do so by referring to the things, which, available in the environment, stand for themselves. By orienting the body, speakers and listeners can provide frames for orienting to and making salient, different moments of the communicative situation. Prosody (pitch, pitch contour, speech intensity, speech rate), too, is an essential communicative element that speakers and listeners draw on for making sense. Words (spoken, written) are parts of communicative situations that refer to the situation in a metonymic (synecdochical) manner, which, inherently, is a one-sided manner that does not express the same that other equally one-sided moments (gesture, prosody, etc.) do.

In this chapter, the conversation involving two scientists talking in the presence of graphs and taking this graph as content is understood to be both *over* and *about* the graph interpretation task. In the latter case, the talk is used to articulate visible features and to make new, heretofore unattended-to features into the perceptual field of the co-participants. In the former case, the talk may not refer directly to a graph feature. Rather, the graph may serve as the backdrop against which the talk occurs and which thereby contextualizes the talk.

ETHNOGRAPHIC BACKGROUND

In this chapter, I take the approach common to other ethnomethodologically and conversation analytically informed studies: Rather than interpreting events from the outside, I am interested in what transaction participants make available to each other as resources to mark and remark the sense of the ongoing task, the context, and the processes. Thus, I do not assume someone to be thinking or seeing something unless he or she *formulates* what is currently happening as such. I assume that the participants produce what they do *for* one another and for themselves *for the purpose of* moving the task ahead and towards its (indefinite) conclusion. What they make available to each other, they also make available to the analyst, who, as vicarious participant, has to have the same communicative competencies that the participants presuppose of each other. The analytical competencies required therefore are of the same type as the (lay) analytical competencies that the participants have brought to the situation. As a trained physicist (MSc) with PhD level work in physical chemistry, I belong, in some ways, to the community of physicists from which the present participants were recruited

The database for this study was produced as an extension of an earlier study focusing on ecologists' interpretation of a set of graphs culled from introductory courses and textbooks in the discipline. For the present study, the larger purpose of which focused on in-field and out-of-field graph interpretations, a structurally equivalent set of graphs was constructed. A total of 21 (mathematical) physicists from one university in Western Canada participated. Seventeen interpreted the six graphs in think-aloud/interview protocols, whereas the other four worked on the tasks in pairs.

The analysis of the sessions with the individuals showed that they focused on different aspects of the graphs. Individual sessions, however, are less than ideal for studying the question of how visibility emerges and becomes publicly available, because the individual does not necessarily think aloud about what is salient or relevant. On the other hand, because pairs of participants are asked to produce one interpretation, they need to ascertain that they are aligned with respect to what they are talking about, what they see or what dawns in their perception, so that they can talk about the feature or assist others in seeing the same.

In this study, I use the beginning of one session to show processes at work by means of which the visibility of features come into being. The same processes can be seen at work throughout the database in those moments, for example, when the research assistant collecting the data is asked to assist the scientists in their interpretive work or asked to provide the sought-for interpretation. In such situation, the assistant did not just tell the authoritative interpretation that the biology professors used to evaluate their undergraduate students, but in fact engaged the participants in a tutoring session during which the scientists were led to identify and interpret particular features.

By the time the two physicists began the interpretation of the graph featured here, they already had completed two others, a line graph showing the relationship of plant densities on elevation (correlated with temperature and moisture) and a population graph. At the moment of the session, Tony had 36 years since his PhD (30 as professor, the remaining as postdoctoral fellow) and Alex had 6 years (5 as postdoctoral fellow). They specialize in astrophysics and the modeling of dynamic (fluid) processes, respectively. They publish between 0.5 and 3 journal articles per year. Both mainly teach undergraduate students in the department. In the think-aloud part of their session with respect to the present task, the two do not arrive at producing the authoritative interpretation.

The task at hand consists of three isographs (Figure 5.2), which means, the lines of constant effect of two nutrients on plant growth are shown for three different scenarios. In the first scenario, the effect of increasing a nutrient is constant for a given value of the other starting with a critical value (at the corners)—one resource always constitutes the limiting resource. In the second scenario, the two resources are exchangeable such that a certain amount of one nutrient is equivalent to a certain amount of the other nutrient—they are additive resources. In the third scenario, the effect of the two nutrients is non-additive with a minimum relative amount required at the apex of the elbow—they are complementary resources.

BRINGING GRAPHS AND FEATURES INTO BEING

In cognitive models, the perception of graph features is purely mechanical, whereby the structures of an external display is "input through low-level perceptual processes" into a "visual buffer" that is part of pictorial

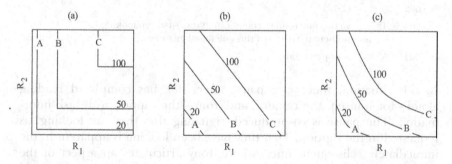

The amoung aplant grows depends on a number of factors, for instance, the availability of nutrients (R). A shortage of any single nutrient can limit plant growth. Sometimes scientists study the effect of pairs of nutrients. The graphs depict three different biologically-realistic scenarios of how two nutrients (R_1 and R_2) might combine to affect plant growth.

Discuss the effects of different levels of the two nutrients on each amount of plant growth (20, 50, 100) in each (a, b, c).

Figure 5.2.   One of three graphs culled from an introductory university-level ecology course that was used in the think-aloud protocols eliciting graph interpretations.

short-term memory or "the mind's eye" (Tabachneck-Schijf et al., 1997, p. 318). In this chapter, I do not dispute that there are visual processes in the mind; I have no evidence about the physiological processes involved while someone interprets a graph. Rather, what really is important to this chapter are the processes by means of which graphical structures come into being as resources for collective reasoning. And this process is more complicated than assumed in the cognitive science literature, because, as previous research shows, "experts" are not necessarily *aware (conscious) of* the height of a line to be interpreted but may in fact be conscious of, and therefore act toward and with, the slope. In this section I describe and theorize how graphical features come into being as two scientists grapple with interpreting one. Because the graph is unfamiliar and not self-evident, they struggle, and in this struggle make available to and for one another everything required to move the task along.

Recognizing and Dealing With a "Busy" Graph

The episode concerning the isograph from ecology begins with Alex reading the caption with task information aloud. After the previous graph, they both turn over to the next sheet and Alex articulates that he will be reading the caption: "Okay, I will read this one, here we go." After he finishes reading (in the omitted turn 001), a pause develops.

002   (2.18)

003   T: so this one is more complicated; s::o:=i=keep looking
             around here to try to put this one together so::=

004   A: =its pretty busy

In this situation, there is a pause after Alex has completed reading aloud—for himself, the camera, and Tony—the caption with task infor- .
mation. The pause is co-produced, signaling that they are looking, as requested in the caption. That they have been looking is apparent in the immediately subsequent lines, where Tony articulates an aspect of the graph, its nature as being "more complicated," which here is understood relative to the four graphs that they have already talked about. Alex provides his own assessment, "it's pretty busy." Tony also articulates for Alex that he "keeps looking around," a description consistent with his prior head movements implementing "looking around" but unavailable to Alex, who had been attending to and reading the caption.

Tony not only has been looking but also, while talking, continues to do so, as he formulates what he is doing. He is not only looking but also doing so with the intent of "put[ting] this one together." The graph(s) are "busy" and "complicated," leaving open the possibility that they are complicated because busy. Because they are complicated, Tony "keep[s] looking around here," where the "here" is unspecified. This process of picking out is not a simple transfer of bits onto the retina and into the visual buffer, but because of the complexity, Tony has to "look around" "to try to put this one together." The busy-ness and complexity further is communicated in the temporality of production, which indexes the searching of keeping looking around, as shown in the following.

```
005              (0.15)
006      T:      yeah; so we got r=ones along one axis and r=twos along
                 the other axis (0.20) .hhhhhhh (0.81) ((total of 1.72)) and
                 we got our twenty fifty hundred lines the:::re=a:: bee ce
                 cases::::=
007      A:      =and it doesnt tell me the units of any of this one. but i
                 guess it do[esnt] matter?
008      T:                 [no; ]
```

There are long pauses, and a long aspiration (.hhhhhhh ≈ 0.7 seconds), all denoting the *effort*-full work required to pick something out from the task sheet. While this is happening, the head continues to oscillate back and forth, from left to right and from right to left. There also is a significant (150% of original) shift in the pitch (97 to 144 Hz) and mean energy (65 to 71 dB, corresponding to quadrupling of speech volume) before voice returns to normal toward the "there." The significance of the feature articulated in speech thereby becomes salient. In this process of orienting toward and "looking around," features are noted and, by means of denotation, brought into collective awareness. Thus, Tony's utterance denotes the presence of two axes, one associated with R_1s, the other associated with R_2s, and he denotes the presence of lines cases. The three lines denoted as 20-, 50-, and 100-lines, being produced with four times the speech intensity of what follows, are hearably the same that are also denoted as "a, bee, ce" cases. Here the fourfold decrease in speech intensity allows us to hear the utterance as a clause.

The two exhibit to one another concurrent attunement to the graph and to each other's talk. The final part of Alex's utterance ("but I guess it doesn't matter") and Tony's confirmation ("no;") affirm that what is visible and the interpretation of its relevance are salient to the two, and that Tony agrees with Alex about the irrelevance of the units. Even though Tony says no, he affirms the assessment, which was stated in the negative. Agreement also could have been signaled using the word "yes," whereas in French the affirmation would have been "si si" rather than "oui." They have noted so far that the graph is busy and begun to collaborate to deal with identifying relevant features of the graph, including the absence of certain aspects such as the units. In noting the absence, they make this feature of the graph stand out; but in noting it *as absence* from this task, they also impute the fact that the absence does not matter to the task asked of them. It is not just that existing perceptual features come to be encoded and put into the "mind's eye" but importantly, that the absence of features is also denoted even though there is no evidence for either Alex or Tony to have produced a mental image of the graphing task with this absent information.

Producing Joint Orientation and Finding Structures

A key condition for collaboration is joint orientation to the same object. If participants in a conversation could not assume that they are oriented toward the same object, addressed as the subject of their utterances, then no communication would be possible. (Communication derives from the Latin *communicare* from *com-/cum-*, with, *munis*, bind, *-are* [-ate], suffix turning adjective to verb.) Orientation begins when the participants turn over the sheet from the previous graph and thereby reveal this graph, which is printed in landscape. Both turn their respective heap of sheets, visibly to and for one another, and thereby indicate how the sheet has to be oriented so that what is potentially visible actually can (and does) become visible. And they exhibit to each other and to the researcher their joint orientation.

```
009             (1.35)
010    A:       discuss the twen- uh: these are different levels:: of: (0.28)
                two nutrien:ts:::
011             (0.52)
012    T:       x=and.
```

In repeating the beginning of the task instruction ("Discuss the") Alex signals his orientation toward the text rather than toward the graph itself. He then makes a statement about "these," where "these" point to what there is, and what it is captured in the concept. The first and second parts of the utterance involve very different intonations, allowing us to hear a reorientation. Thus, prior to the break the mean intensity of pitch (F0), F1, and F2 were 66 dB, 102 Hz, 665 Hz, and 1811 Hz, whereas for the second part the corresponding values were 70 dB, 89 Hz, 464 Hz, and 1456 Hz. In the second instance, we can hear the utterance as a realization, even surprise, whereas the former constitutes a "mere" re-reading of the text.

There is an element of a surprise, "uh," and then a statement that "these" "are different levels of two nutrients." Stating it here makes this salient. Making the statement is for a purpose, as stating what goes without saying requires an explanation. Why would someone say "I am walking" while walking, which is evident to everybody in the situation—unless it might be a telephone conversation where the recipient cannot see the speaker? And then, it still requires an explication or some form of rationale for stating something such mundane. The utterance "these are different levels of two nutrients" therefore constitutes both a *description* of and an *instruction* for how to look at the display in front of them.

A similar situation follows immediately after, this time it is Tony who describes and instructs Alex what to see in or how to look at the display.

016	T:	okay, so I dont know; I guess if you pick a scenario like (0.56) be::; to pick one in the middle (1.56) as as = uh::::::::::
017		(1.59)
018		how does this work. lets see; so this (2.84) ((drags the sheet of graph in front of himself)) ↑OH you know; ((79dB, 135 Hz))
019		(1.97)
020	A:	uh:: (0.20) two nutrients; r=one=were plotting.
021		(0.50)
022	T:	OH, THESE ARE CONTOUR ((147 Hz)) this must be contour lines. right? so this is (0.47) .hhhhhhhhh (0.49) these twenty fifty hundred must be contour lines
023		(0.22)
024	A:	ya about; ya. ((points to last line of the caption))

Tony orients Alex and himself to the "bee" (B) curve "in the middle," the middle one of the three panels (the voice parameters change so that we can hear this part as a clause explaining the choice of "b"). Then he formulates to be looking "Let's see," and he does so more closely, as indicated by his pulling of the sheet closer to him. And then he expresses surprise ("OH"). Tony announces something Alex should know, "OH, you know," but there is a pause without Tony revealing what it is that surprises, it may even be emergent for himself, and then he repeats the interjection that expresses surprise. The emphasis is on the "these" and "contour" produced by significantly higher than normal pitch (average is 147 Hz, peaks at 165, 187, 163, 176, 208 Hz against a background pitch of below 100 Hz).

There are multiple "B" curves, and Tony directs Alex to the one that is part of the "scenario" in the middle. Clearly, pulling the sheet closer to him is his response to his own question, "How does this work?" To find out, he needs to be closer to the sheet allowing him to more closely inspect it. He announces the question he poses both for Alex and himself, and then engages in the endeavor to find an answer. That is, he describes (formulates) the work that he is in the process, both now and prospectively, of being engaged in.

These are the singular lines visibly available to them, there are several, and lines; all of which allow the listener to pick out what the physical thing is that is classified as "contour lines." These lines no longer are singular but become realizations of something more general, "contour lines." The "twenty" "fifty" and "hundred" denote the things that are classified as contour lines. They are thereby made visible, and Tony simultaneously orients Alex to the things to which the terms twenty, fifty, and hundred can be assigned. But the visible lines, letters, and signs materially embodied in the sheet before them make salient certain aspects of the sound stream. Thus, the sound we hear as "twenty" becomes salient because of the co-presence of the "20" on the sheet of paper, which culturally historically we have come to perceive (hear/see) as alternative expressions of the same "idea." The utterance of twenty (as that of fifty and hundred) makes salient structures in the complex of black ink and paper; and the complexes of ink and paper make salient aspects of the sound stream. This sound stream provides further structure, which is part of the marking and remarking of structure, which is part of marking and remarking sense. A bodily and embodied way of producing covariant structures that alone mark salience, which has to be available for the mutual alignment. In this situation, therefore, the two are both aligned and produce alignment. Without existing alignment and intersubjectivity, they would not be able to produce further alignment and intersubjectivity.

Alex orients Tony and the onlooker to the text, his hand being placed[2] and by re-reading a part of the caption. The text appears to be the same text again, but, precisely because it is a reproduction of the text, it no longer is the same (Bakhtin, 1986): the talk pointing to the problematic nature of the text. Alex makes visible that there is a problem in the text, which specifies something about what can be seen or how what can be seen is to be interpreted (to be seen). He is thereby alerting us that he is oriented to the visible text. Alex produces signs of agreement, "Ya, about ya." He then orients himself and Tony toward the caption, placing his hand next to it, and reading—again—a part of it.

```
025     (0.40)
026  T: is that what they are?
027  A: nuh not really. well, these are, oh well, me– (0.20) i guess
        they are in a sense
028     (0.98)
```

Figure 5.3.   The hand orients Alex toward a particular part of the text, and his listener to the fact that the utterance reproduces a reading of the caption.

Tony himself offers what can be heard as a query both grammatically—he uses the pronominal "what"—and a rising pitch toward the end, heard as a query: "Is that what they are?" With Alex's utterance "Nuh, not really," the two turns are realized as a question–answer sequence. The answer is tentatively negative, "not really." There is a lot of hesitation, and then Alex proffers a description that leaves the possibility that the lines possibly ("in a sense") are contour lines. There is no confirmation, only the possibility that the lines are contours, leaving the ontological status of what there is still open.

```
028        (0.98)
029    T:  so; in other words:[::  ]
030    A:               <<p>[the] graphs are three different biologically real->
                for ´two nutrients, these are ^nutrients
031        (0.71)
```

Tony attempts to reformulate, "so in other words," but discontinues as Alex re-reads once again a part of the caption. But now, in contrast to the first time, he strongly emphasizes the words "two" and, in the second occasion, the "nu" of the "nutrients." Because of the prosodic emphasis, the semantic content of the sounds "two" and "nutrients" are made salient. This provides a structure for finding perceptually, in the display, something that is equivalent. It is only when there is an equivalence that what is described verbally and what is available perceptually constitute an objective presence. The utterance then is a description and an instruction for looking at the display in a particular way, namely for identifying "two" things; and when there are two things perceptually available, the two levels of entities reify one another.

Alex begins, talking under his breath as if for himself; and yet, that fact that he is audible signals to others what he is doing. As he comes to the end of the sentence he currently is reading, his voice intensity moves into the normal range, and then he repeats the word "nutrients." Here, the term "nutrients" is made salient in the repetition, the elongated production (0.98 s) (compared to "these are" [0.21 s]; 0.50 s in first instance).

The graph, and the possible forms of sense it marks out, slowly emerges as features become and are made visible. We witness a movement from a "busy" and "complex" display to one that has specific lines that can be seen as representing something not actually seen, in a third dimension, made available through hand movements that show a plane rising from front (close to Tony) to the back (away from Tony).

In the following, Tony makes salient the possibility that the lines are "contour lines." He provides his description/instruction by beginning to orient Alex (and the viewers) to the "hundred zone," which one can find along "that line." What that line is has not yet been specified, but is so with the next utterance in which he further elaborates the "meaning" of the previous. He places his right index finger on a point far to the right and near the line named "20 (A)"; and his left index finger comes to rest on the ordinate corresponding to the ordinate value of the point (Figure 5.4a). Then he orients us to a point on the same line but on the other side, placing the two index fingers in the corresponding points low on the abscissa but high on the ordinate (Figure 5.4b). Again, the placement of the two index fingers and the particulars of the display mutually elaborate (motivate) each other, because the placements are seen to be intentional.

In fact, he does not orient us to a line but to a zone. The zone means something, here that one can have different combinations (he uses the inclusive "or") of R_1 and R_2 that are in the same ("hundred") zone, that is, the line corresponds to something that stays the same. Tony then articulates his conclusion, "It must be … a contour." There are three repeated attempts in producing this naming, which is accompanied by a hand movement that can be seen as an iconic representation of an inclined plane or, as he then articulates, a hill.

031   (0.71)

032   T: s:o youre in the hundred zo:ne; (1.99) <<f>so if you go to the middle of gra:ph>. youre in the hundred zone (.) down that line; which means that you can have a lot ((points on the graph with two fingers Figure 5.4a)) of r=one and a little bit of r=two: ((points on the graph with two fingers Figure 5.4b)) or lot of r-two and a little bit of r=one, it must be like con-; it must be contour like uh=, ((repeatedly gestures up in the diagonal direction with right hand Figure 5.4c, Figure 5.4d))

033   A: =ya
034   (1.18)
035   T: a contour .hh sa, uh::: like a hill there
036   A: uh hm,

In this, the hand with the open palm moves over the diagram, along the diagonal of the three curves, away from Tony (Figure 5.4c, d). Then hand ascends as it moves across the lines with larger numbers. "It must be ... like a hill there," and the "there" coincides with the hand movement over the three isoclines, instructing us to see, or articulating the possibility for, a hill being there above the diagram, somehow indexed by the diagram but not itself visible.

Here, the hand makes visible for us something that somehow is contained in the figure but not actually perceptually available. It is pointed to but has to be envisioned to be seen. The hand instructs us what we have to envision, a part of a hill, which has a side that parallels the three lines, and which ascends with ascending numbers on the lines below the hand. Tony describes what can be seen, and this description is available in his hand; it is also an instruction for how to look at the graph and what to

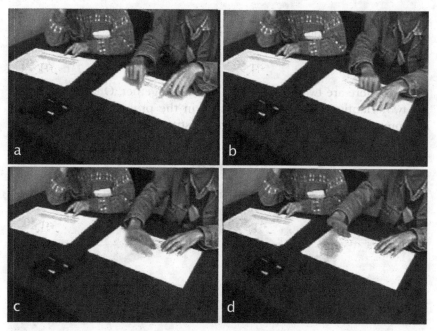

Figure 5.4. Gesturing orients both speaker and listener and, in a dialectic relation with the material underneath, highlights structures thereby made available in the graph.

create, and the precise hand placement instructs us about the relationship between the placement and orientation of the hand—the sides of the hill—and the inscription below.

We can see three beginnings of the utterance "contour," but the hand does something that is not iconic with a contour line. Rather, the hand is iconic to the side of a hill, a description that follows. There is therefore a contradiction between the utterances of contour and the visually available movement and orientation of the hand. This contradiction finds its resolution when the naming changes to "hill." Such contradictions have been described to occur when people are engaged in exploratory talk, new to a domain, and when individuals are attempting to develop ways of talking about phenomena or talking phenomena into being (Roth, 2005). This extension of the graph into a third dimension can also be seen in the following excerpt from the session, in which Alex uses his hands to make visible a third dimension orthogonal to the two spanned by the axes on paper (Figure 5.5). The offprint, which overlays three images simultaneously, shows how the hand moved up and down to produce a perceptual representation of this "third axis" that he also describes in words (turn 049).

```
048        (2.63)
049        thats the third axis in other words.
050        (1.17)
051   A:   yah.
```

Here, there are two levels of talk blended together. On the one hand, the two scientists talk about the graph in the process of producing an interpretation. But the graph also constitutes a background against

Figure 5.5.   A third axis comes into being, made visible in the hand movement from the graph on paper vertically upward, literally producing a third dimension, even though ephemerally.

which talk and gestures become salient. The paper spans two dimensions parallel to the tabletop and the task paper. It is against the existing drawing that lies flat on the table that the hand gesture moves vertically up and down and thereby produces the image of the (ephemeral) third dimension and axis. This hand gesture thereby makes salient that what can be imagined but that is not there materially other than in the ephemeral gestures. Other gestures with or without material support, such as a pen, make salient physical features in the graph. For example, in the following excerpt, Alex develops a narrative account related to the middle graph. He begins to articulate something about a slope of 45 degrees and then suggests that these "are" (i.e., mean) equivalent nutrients (turns 117, 120). Precisely at the end of uttering "equivalent," his pen drops at the first of the three lines (Figure 5.6a) and then, while uttering "nutrients" an instant later, pulls the pen across all the lines on the sheet of paper.

```
117   A:   so here ((b)) this is almost like a f (0.23) yea fortyfi:ve or whatever the
           slope is increasing; an these are basically equivalent- ((places pen, paral-
           lel to the lines, [Figure 5.6a]))
118        (0.37)
119   T:   ye[ah].
120   A:   [nu]trients                                    ]
           [((sweeps pen across the three lines))]
121        (0.21)
```

Here, an aspect of the graph becomes salient in the dialectical tension that exists between the pen, its direction, and the direction of its movement across the paper. On the one hand, something in the graph motivates the particular orientation of the pen; and on the other hand, the

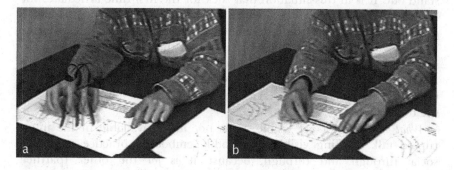

Figure 5.6. Gesturing orients both speaker and listener and, in a dialectic relation with the material underneath, highlights structures thereby made available in the graph.

pen motivates the process of finding (seeing, perceiving) something in the graph that corresponds to it. Thus, it is not merely that there is something highlighted by the pen, but there is a mutually constitutive relationship between the pen and its orientation, on the one hand, and the objectivity of graphical features, on the other. This is also the case for the moving gesture, which, while holding the slope of the pen constant and thereby parallel to the three lines, features something that is constant across the movement. Here, equivalence of the two nutrients pertains not only to the three lines but to all the lines that are not drawn but made available in the sweeping motion that makes available lines wherever the pen moves. In all of these cases, the gestures constitute not only descriptions of what there is to be seen—realized in the perception of the speaker—but also instructions for how to find something to be seen.

The gestures—with and without additional materials in the hands ("tools")—do not only foster joint orientation and highlighting of physical features, but also the interpretation of the features thus detected. For example, throughout the following turn, Alex holds the pen such that it covers the horizontal branch of the graph denoted by the letter "A" and the number "20" (Figure 5.6b). This gesture makes salient the bottom branch rather than the curve as a whole. It is at this moment that Alex articulates for the first time the independence of the curve with respect to the variable R_1.

127   A: [the r=one:: a:h (0.65) independent of r=one]
          [((pen is placed on a horizontal branch of the 20 curve [Figure 5.7b]

In this second example (Figure 8.6b, turn 127), the pen lies on one of the lines in the first of the three graphs. The iconic relationship between the graph, a horizontal line, and the pen above it, makes the horizontality stand out. It is in this situation that Alex for the first time articulates the "independence" of the phenomenon with respect to the nutrient R_1 *in this part of the graph*. Here the independence emerges from the description. There is no evidence for the conjecture that the "independence" of the strength from R_2 is given, salient, known. There is hesitation, a long gaze at the pen, which brings out the salience of the feature under investigation. The feature is emphasized as well as made visible to the other and to the researcher.

What it is, is not yet visible, or is not made available in the interrupted talk. Making visible is a bodily embodied practice, and it is social through and through, because it is *for* the other (partner, researcher), presupposing intelligibility of the utterances and gestures, drawing on resources that have come from the other (language). What matters is not the raw stimulus: it is important to perceive, whatever there is, as an instance of something, which means, that the visibility implies the bringing together of the singular, *this* graph, with *these* features, in the here and *now*, and the general, this graph as a concrete realization of some concept or category.

The Visibility of Mathematical Action and Reasoning: Formulating

The participants not only make visible features available in and through their communicative efforts—talking, iconic gesturing, orienting, pointing (gesturally, prosodically)—but also make visible aspects of the event itself, such as when the participants formulate what they are doing in words or gesture. That is, to derive a collaborative interpretation, the participants need to articulate not only parts of the graph and what they mean but also the have to articulate the various mathematical processes that they enact. Formulating therefore allows participants not only to do mathematics, but also, simultaneously, to articulate and name what they are doing and why they are doing it without leaving it to the listener to impute or infer any hidden intentions. This is a central aspect of research design in ethnomethodology, where participants are either put into a situation where their normal ways of accomplishing a task no longer function (e.g., using inverted lenses) or followed into situation where there are "natural" breakdowns in their everyday routines (e.g., observing someone cooking, using a field guide, or doing a science lab experiment for a first time). In the following utterance, Tony makes available to his partner, the research assistant, and the camera that what he is looking at is "more complicated" (than what they had looked at before). He then *formulates* what he has done and what he keeps on doing: "looking around" (turn 003).

003   T: so this one is more complicated; s::o:=i=keep looking around here to try
              to put this one together so::=

As this instance shows, we do not need to make hypotheses about the private intentions of the person but rather, Tony makes available to everybody who is listening that he is looking around. More so, he "keep[s]" looking around, which indicates that he *has been* looking around, the result of which he has articulated as an object ("this one") that is more complicated. He also formulates the goal of his looking around, namely to "try to put this one together."

In some instances, the denotation of the mathematical actions is instantiated in a more oblique manner. In suggesting that "it doesn't tell me the units of any of this one," Alex formulates that he has been reading and looking for units but has not been able to find them (turn 007). In fact, he attributes the problem to the text, which does not tell him something he needs rather than that one related to his competencies of being able to find what is there.

> 007 A: =and it doesnt tell me the units of any of this one. but i guess it do[esnt] matter?

He does not simply say that it does not matter but he formulates, for Tony, the researcher, and the camera, that he makes a guess about whether the information matters. That is, he also formulates uncertainty about the requirement of the feature he has been looking for. There are still other means of indicating what one is doing. Thus, when Alex utters the text from the caption again, he thereby points to the fact that he is reading:

> 009   (1.35)
> 010 A: discuss the twen- uh: these are different levels:: of: (0.28) two nutrien:ts:::
> 011   (0.52)

Here, Alex is not merely reading, but also, because we can hear this reading as a re-reading, the meaning changes. Because actions are taken to be motivated, re-reading a text that has been read before also suggests that there is a need for re-reading, which may come from not having understood what the text is saying or from having forgotten what it is saying. In the following, Tony articulates not knowing something. He then articulates making a guess about the next step ("pick a scenario"), and that he is picking the one he does for a particular reason ("pick one in the middle").

> 016 T: okay, so I dont know; I guess if you pick a scenario like (0.56) be::; to pick one in the middle (1.56) as as=uh::::::::::
> 017   (1.59)
> 018 T: how does this work. lets see; so this (2.84) ((drags the sheet of graph in front of himself)) ?OH you know; ((79dB, 135 Hz))

There therefore is a dialectic of description and instruction at work. Tony says "if you pick," he articulates what he is (has been) doing but at the same time he instructs Alex what to do for moving on to the next yet-to-be articulated step in the reasoning.

In the following episode, Alex articulates by very different means that he is reading—once again. First, his left index finger moves from left to right (his perspective) as he utters the words of one sentence; when he repeats the sequence of words, his finger already has moved to the left again and then moves to the right as he speaks. But this time, he is barely audible. He thereby indicates that he is reading again, and it is because it is again, he does so inaudibly all the while showing us (visually, audibly) that he is reading and what he is reading. With the new material (information), the voice intensity rises back until reaching normal levels. It is precisely on the "so," a conjunctive with causal implication; this implication, which is new, emerges as the utterance unfolds: "I guess the biggest number they got."

037   (1.41)
038   A: so is this percent?
039   (0.23)
040   T: no no [no]
041   A: [are] <<dim>r=two nutrients of each amount>; <<p>discuss
           the ^different levels; (0.22) discuss the effects ((the little
           finger moves left to right, twice, precisely with the
           "discuss")) <<pp> of different levels of two nutrients on the
           amount of> <<cresc>plant growth> so ((places hand on b, Figure 5.7.))
           this is plant growth so, .hhhhhhh (0.47) i guess it this
           is thuh:: ((taps, then circles around))), ((follows the 100
           line in b with thumb))) i guess the biggest number they got so.

It is in his body that Alex shows us the connection between a particular part of the text and a particular part of the graph (Figure 5.7). But the implication does not appear to be complete, as he utters the same implicative conjunctive again, "I guess, the biggest number they got, so …"

In summary, this section shows how participants not only align each other to particular features but articulate for each other what they do and how they do it, their sense of certainty about the statements they make, and the attribution of (momentary) problems (e.g., the text, the graph, oneself). Thus, graphing and graphical features are not only about perceiving unambiguous, physical features and interpreting them but also about bringing features into the discourse and providing accounts for the processes by means of which this bringing into discourse and therefore sharing the feature is achieved. Perception is not relegated to physiological processes that lead to the projection of the material world onto the retina and the subsequent processing in mind, but there are processes at work that are essentially social that are required to bring anything into the realm of shared discourse, knowledge, and experience. In the transition between the singular impressions that are formed on the retina and the shareable knowledge arising from the experience of being oriented in and to the world, uncertainty plays an important role in the production and reproduction of a common sense.

Figure 5.7.   The two simultaneous gestures, the left hand toward the text and the right hand toward one of the graphs produce an embodied link between text and image.

Uncertainty

Perceiving / making salient a physical (material) trace in itself is insufficient to bring order to an observation. There are many traces; and there are different ways in interpreting these traces, that is, linking them up with a network of significance. It is therefore only when some visible trace can be seen as a concrete realization of some category that sense has come to be marked and re-marked. (Re-marked, because the feature becomes only a realization of a category that already exists.) More so, the task asks the participants to arrive at an interpretation, which they do not know at the beginning what it will be. It is immediately evident that they cannot have a set of criteria for determining whether what they do leads to any desired useable but yet-to-be known result. It is an age-old problem captured in the concept of the learning paradox or, in a different context, of the gap between plans and situated actions. In the transcripts from the sessions with the physicists interpreting graphs, uncertainty is a salient feature. Uncertainty here pertains to the fact, among others, that participants cannot know whether what they perceive and do is required for the interpretation that they ultimately achieve. In the following episode, Tony first formulates not knowing and then providing a possible scenario for approaching the task ("I guess if you ..." [turn 016]).

016   T: okay, so I dont know; I guess if you pick a scenario like
           (0.56) be::; to pick one in the middle (1.56) as as=uh::::::::::
017   (1.59)

018   T: how does this work. lets see; so this (2.84) ((drags the sheet
            of graph in front of himself)) ?OH you know; ((79dB, 135 Hz))
019   (1.97)

After uttering the premise of "pick[ing] one in the middle," which is followed by pauses and filler utterances, he continues by formulating uncertainty, "How does this work?" The hesitation in continuing is formulated
obliquely as uncertainty about what the implications are that follow from
the premise. Tony then formulates orienting ("let's see" [turn 018]), pulls
the graph up closer toward him. After a lengthening pause, he then
expresses surprise ("OH"), a realization, and then announces something
as knowable ("you know"). In this, some uncertainty has been removed, as
the "you know" is in contrast to the previous description of "I don't know"
(turn 016).

Uncertainty arises because, unlike constructivists ascertain, there are
essential passive dimensions to cognition generally and perception specifically. Thus, we do not know why we see what we see rather than something else; and readers will have had the experience that after passing
some spot for weeks, months, or years they see something never seen
before. The things that appear to us in our perception therefore have as
an essential dimension their "givenness" (Marion, 2002). There are phenomena (perceptual aspects) that give themselves. Precisely because they
are unintended, they may give rise to an uneasiness, without the person
being able to tell in so many words why this uneasiness exists and where it
stems from. There is a contradiction between what makes sense, and other
things that "bother" Alex. These troublesome "ones" are those where R_1
increases, and then the articulation stops.

089   A: well this one here is ((points to "a"))) kind of uh: I guess
           (0.39) [when both but when   ]
090   T: <<p>[this one is weird yea>]
091   A: both ((Figure 5.1)) increase ((left hand moves upward on abscissa))
           ((circles 100 "plane"))
092   (0.96)
093   T: ye[a:]
094   A: [that] makes sense but the ones ((c)) kind of bother me
              when r: when r=one increa:ses:;

Hesitation and uncertainty is expressed in multiple ways. Following
each of the three utterances of "so," there are signs of uncertainty. There
are pauses, together with long audible in-breaths (as in laborious processes), and extended phonemes ("thuh::"). He formulates to be guessing
twice, the second time that "100" (which he points to) is "the largest

number they got," which is heard against Tony's rejection of his assumption that the numbers represent percent. So from his initial perception of 100 as being 100%, he now sees it as a number, and the highest one that an unspecified "they" had measured or produced. Here, the ontological status of the line itself or of the ink trace marking 100 does not appear to be at stake but what these traces denote in the world he/they understand. The question is not about "these" but concerns what "this" or "these" "are" initially, so the possibility is offered that the observation categorical could be completed by "is percent." It is one way of bringing order to the complicated and busy display. In fact, it is not the entire display that is reigned to order in this way but one particular aspect, the lines labeled with letters (A, B, C) and numbers (20, 50, 100). It is the ascription to a particular order that allows order to enter the business of the display and to bring order to something that is complicated.

Tony then takes another shot at articulating a description of what is there / instruction for how to look. He also articulates that "he is getting comfortable with the idea," which, assuming that human beings do not simply make non-significant utterances, marks for us that there is a process of becoming comfortable. Comfort contrasts possible discomfort arising from uncertainty. The earlier hesitation therefore comes to be in relief, an initial and yet uncomfortable way of seeing what cannot be directly seen. There was a contradiction between the inclined plane of the hillside visible in his hand and the contour category made salient verbally. He begins by making the same waving movement associated with the term "hill" before. This articulation therefore connects his previous turn to the earlier one in which he first produced the gesture and word. Then he produces what we can hear as a verbal ("as you move around here") and gestural elaboration (moving his finger along the contour line).

He is not moving around anywhere, but around the "R_1-R_2 plane," which clearly sets this graph apart from the four previous ones he has interpreted in which the abscissa denoted an independent variable and the ordinate, a dependent variable. Now he articulates a plane spanned by the two variables and he articulates the lines as "different altitudes." While he articulates the verbal description (instruction what to look at), he places his hand with palm parallel to the paper, raises it, and returns to the lower position; he does this three times. He then suggests that "these things," which we hear and see to be the three different lines as his hand sweeps across the center plate, as "indicating the altitudes." Thus, he first produces a visual image of planes at different "altitudes," and then, by repeating the term "altitude," links "these things," which we observe to be the lines as the hand swipes over them, to the image of the horizontal plane that he had produced with his hand. It is in the repetition of the term "altitude" that the link between the visually available horizontal planes and the *subsequently* visually denoted (indexical gesture) lines comes to be established. The "here" points to the particular display underneath his hand. Tony's talk is pedagogical, as he teaches us how we have to look at the graph and what we have to envision, and how the perceptually available and envisioned come to be connected. There is a considerable pause, and then Tony reformulates and therefore elaborates ("in other words") that "that's the third axis." What the "that" is remains unspecified and the listener has to find it in the context, with a great likelihood being that it pertains to the content of the immediately preceding communicative action, that is, the altitudes.

TALKING AND SEEING

In this chapter, I draw on the example of two scientists engaged in a graph interpretation task to articulate essential elements in bringing the visibility of graphing features into the discourse and, therefore, in making these features part of a shared world. These features are prevalent throughout the database and particularly when pairs of scientists are involved, as in this case the task is one of arriving at a collaborative answer to the task rather than verbalizing aloud private thought. In the process of engaging with the tasks of interpreting the graphs provided, scientists articulate for one another what they are doing, what there is to be seen, and what there is to be concluded. In being articulated, however, every resource that is highlighted in and through denotation is objectively available to the other generally. Anything that is there and can be denoted by language or some other sign is present objectively in the realm of the ideological. It is not something ephemeral, in some transcendent realm of being, but available in and as of concrete material reality.

To become discoursable, visibility, as structure in the display, has to be shared. Visibility has to be intelligible, and has to lend itself to be shared and accounted for. But discourse also is used to bring visibility about. Visibility and discoursability are intertwined and reflexively related. Because visibility is not given, because it emerges from the transaction, it cannot be intended and therefore constructed like one intends to construct a house. It is in and through the collaborative conversation that aspects become visible, discoursable, and available to practical consciousness. Communication generally and language in particular, which is "practical-consciousness-for-others and, consequently, consciousness-for-myself," is both the means and the ground for establishing the *shared* visibility of graphing features. The episodes presented here show that the visible aspects are not simply there: aspects dawn and then are confirmed and solidified as communicative forms (talk, gestures) around them develop. Thus, the hillside was not something immediately apparent to Tony. Rather, something seems to dawn to Tony, but it is coming forth only with and in time. The aspect is not there, that is, neither the hill nor the layered representation that yields the isocline plot. It is an already theoretically laden image the projection of which is available at hand.

Throughout the episodes, alone or working with a partner, the scientists used gestures to point to, mark, iconically represent, animate, create, and make visible graphical features. What there is, the structured display comes *to hand*, and it does so in a progressive manner. Coming to hand here is meant both literally and metaphorically. Literally, the participants point to and touch and produce gestures that have perceptual similarity with entities displayed. This happens progressively, because the display was "busy" and "complex" initially, and gains clarity in the transactional process.

Finally, seeing and perceiving are not instantaneous processes whereby an image of the external world is created in the mind's eye. Rather, the existence of thoughts, models, concepts comes into being in real time, through incremental processes, which develop and concretize what there is that can be interpreted. In some instances, this may be the case, such as when Tony, after not having spoken for a while, announces that he is "getting comfortable" with some idea. In other instances, there is no time-out for one or the other to reflect, so that the public articulation and the thinking are inseparable. In any event, there is a difference between having some idea (model) and developing it with specific details. The model becomes real for the speaker and the listener simultaneously.

ACKNOWLEDGMENTS

The work reported in this chapter was made possible by a grant from the Social Sciences and Humanities Research Council of Canada.

NOTE

1. This is the aspect for which the researchers are responsible. Readers find more about this form of analysis in Roth and Middleton (2006).

REFERENCES

Bakhtin, M. M. (1986). *Speech genres and other late essays*. Austin: University of Texas Press.

Marion, J.-L. (2002). *Being given: Toward a phenomenology of givenness*. Stanford, CA: Stanford University Press.

National Council of Teachers of Mathematics (NCTM). (2000). *Principles and standards for school mathematics*. Reston, VA: Author.

National Academy of Science (NAS). (1996). *National science education standards*. Washington, DC: Author.

Roth, W.-M. (2003). Competent workplace mathematics: How signs become transparent in use. *International Journal of Computers for Mathematical Learning, 8*(3), 161–189.

Roth, W.-M. (2005). *Talking science: Language and learning in science classrooms*. Lanham, MD: Rowman & Littlefield.

Roth, W.-M., & Bowen, G. M. (2001). Professionals read graphs: A semiotic analysis. *Journal for Research in Mathematics Education, 32*, 159–194.

Roth, W.-M., & Bowen, G. M. (2003). When are graphs ten thousand words worth? An expert/expert study. *Cognition and Instruction, 21*, 429–473.

Roth, W.-M., & Middleton, D. (2006). The making of asymmetries of knowing, identity, and accountability in the sequential organization of graph interpretation. *Cultural Studies of Science Education, 1*, 11–81.

Tabachneck-Schijf, H. J. M., Leonardo, A. M., & Simon, H. A. (1997). CaMeRa: A computational model for multiple representations. *Cognitive Science, 21*, 305–350.

Vygotsky, L. S. (1986). *Thought and language*. Cambridge, MA: MIT Press.

Wittgenstein, L. (1958). *Philosophical investigations* (3rd ed.). New York: Macmillan.

EDITOR'S SECTION COMMENTARY

Wolff-Michael Roth

In the preceding four chapters, the authors present us with interesting case studies of people dealing with transformations of various kinds. Edwards describes for us what happens when teenaged students engage with a computer program that allows them to enact geometrical transformations on a two-dimensional plane. The transformations include the commands "slide," "rotate," and "reflect" that the students use as part of tasks or "games." The author notes that unlike mathematicians, who see geometric transformations as mappings of the plane onto itself, students understand what they do as moving objects about the surface. From a purely experiential perspective, given the resources that the students have (the commands *slide*, *rotate*, and *reflect*), we might have expected students—if they consciously thought about their actions at all—to experience what they are doing as moving the given objects using the given commands. Perceptually, the do not map the plane onto itself, for experientially, the plane remains constant; perceptually, they move the objects. If the idea is to allow students to map a plane onto itself *and* allow students to draw on their embodied experience of the world, the mapping process itself may have to become more salient in the task. For example, a

Mathematical Representation at the Interface of Body and Culture, pp. 123–127

teacher or researcher may provide both source and target domains in different layers, then ask students to transform one layer so that it is identical to the other. In this case, not only the object but also the domain as a whole undergoes the mapping (onto itself). In this case, the embodied experience in and with the world mediates access to the mathematical understanding of geometrical transformation.

From a cultural-historical activity theoretic perspective, students act in a (life-) world given to them; it is their starting point because it is the only place in which they can act. Any development of their practical understanding and consciousness about what is happening in the transformation, from the perspective I take, *follows* practical engagement with the world to the point that it is familiar and then can be reflected about. This development parallels that of the cultural-historical development in the construction of cathedrals. Thus, there were first masons and master masons, who, with the simplest of tools, constructed giant cathedrals in a bricolage fashion until a point where the cathedrals became so big that the master masons felt a need to formalize the construction process. At this point, they transformed into architects and no longer built themselves, instead telling ordinary masons what to do according to the plans that the newly minted architects have drawn on paper. Here, too, consciousness for what they were doing emerged for the master-masons-turned-architects only after having had tremendous experience with building cathedrals until the inner contradictions in their phenomenal approach no longer sufficed to attain the size of building they were aiming at. Thus, I would expect Edwards' students to develop a sense of transformations as mappings of the plane onto itself only *after* having practical experiences thereof.

In a similar way, the students in chapter 3 "struggle" with the tension between a phenomenological reading of graph segments in relation to a fixed spatial point, on the one hand, and a reading that puts emphasis on relative relations, on the other hand. Radford articulates the similarities between the slow and difficult historical evolution of a Cartesian representation and the teacher-mediated Cartesian treatment in his tenth-grade classroom. In this development, students' being-in-the-world itself became a resource: They make use of a polyphony of voices, gestures, and mathematical signs. The new understandings arise in and from the practical engagement with the task, so that the task itself provides the conditions for the new understandings to emerge. Thus, as cultural-historical activity theorists emphasize, it is the participation in concrete practical and collectively motivated work—here related to mathematical inscriptions—that new forms of consciousness emerge, here as practical understanding of story content and its representation in a graph. Practical engagement therefore is leading

development rather than development preceding practical engagement. Thus, the students' practical understanding of the world constitutes the ground from and upon which new practical understandings
develop in response to the inner contradictions that become apparent,
for example, when students realize that in their explanation an explicit
presence of Marthe is missing.

In chapter 4, second-grade students are asked to find in their classroom other objects that are similar to the one that they have received
from the teacher. Maheux and colleagues articulate for us the way in
which cubes are appropriated in and as cultural objects in a simultane-
ously occurring double ascension of the abstract|concrete dialectic—from
abstract to concrete and from concrete to abstract—an articulation that is
entirely situated in cultural-historical activity theory. Thus, development
always moves from a "dim stirring," which constitutes the beginning of
thought like a seed, to its full entirely concretized formulation (Vygotsky,
1986). Two processes in fact are at work—thinking and communicating—
and there is a continual (transverse) movement back and forth between
the two processes allowing the relationship between the to develop itself.
Through this relationship evolves the longitudinal movement of ontogenetic development (development in a functional sense). Vygotsky, though
speaking about the word rather than communication as a whole, anticipated the movement from abstract (undifferentiated whole) to concrete
(differentiated whole) and the simultaneous development from concrete
(word) to abstract (idea):

> Semantically, the child starts from the whole, from a meaningful complex,
> and only later begins to master the separate semantic units, the meanings of
> words, and to divide his formerly undifferentiated thought into those units.
> The external and the semantic aspects of speech develop in opposite direc-
> tions—one from the particular to the whole, from word to sentence, and the
> other from the whole to the particular, from sentence to word. (p. 219)

The production of the individual word, which, in everyday speaking, we
do not select but which emerges in and from our mouths, is an operation,
whereas the completed sentence (or utterance) constitutes the speech act.
Vygotsky's approach, therefore allows us to see that the utterance (action)
is an *emergent* product that realizes a goal (idea) that does not have to be
completely formed prior to the beginning of the action. Rather, this form
of acting in the world resembles the way bricoleurs proceed in and with
their work: They begin only to question afterwards how to continue and
find out more precisely what want to do. In this, they are in the same situation as scientists, who often find out long after the fact that they had not
done what they wanted to do, even when the tasks are as simple as a

research biologist doing a dissection no more difficulty than the ones that high school students do (Roth, 2008).

In the final chapter 5 of this first section, I describe how in a collaborative task of graph interpretation, perception itself becomes problematic. The chapter shows that the visibility of lines and other constituents of a graphical diagram cannot be taken for granted but is an emergent property of collective perceptual processes, which, in this case, the participants articulate for one another as part of a joint project. The chapter also shows that visibility is not a question of interpretation, a process that can only begin once some material trace (sign) is given in consciousness and understood as referring to something else. The chapter is more fundamentally concerned with the preconscious and unconscious processes and dispositions by means of which lines and other traces become not only visible but also *accountably* so, that is, as collective and therefore social properties. This latter part is important, for in communication with others it does not suffice to claim that something (trace, sign) exists but it has to exist *objectively* for the participants specifically and for any other member of the relevant cultural community more generally. That is, here again we see how the fundamentally unconscious operations are tied to inherently social processes because of the central position of actions. In a cultural-historical activity theoretic approach, embodied cognition and social cognition are not two different approaches: they are two moments of the same approach.

Across the four chapters, we see a development whereby the articulation of the embodied and social aspects of mathematical cognition is increasingly made visible. In Edwards' chapter, the students are not aware of and do not have the resources to move from their embodied experiences toward the formal ways in which mathematicians approach the issue of transformations of the plane into itself. Radford shows us how, by means of appropriate teacher mediation, students are provided with resources that enable them to move from phenomenological to Galilean ways of understanding the graphical representation. Maheux et al. show in the very act of communicating, children find themselves in a situation where their singular and unarticulated bodily experiences are moved, by means of the requirement to communicate, into the social realm that enforces (shared) intelligibility. Finally, my own ethnomethodological analyses of physicists asked to interpret graphs show that what might be expected to be merely embodied perceptual phenomena in fact already are utter social phenomena. Cultural-historical activity theory, with its dual attention to the embodied and the social, provides the means for understanding not only each phenomenon in isolation, but also the integration of the two into a coherent framework.

REFERENCES

Roth, W. -M. (2008). Radical uncertainty in scientific discovery work. *Science, Technology & Human Values* doi:10.1177/0162243907309627
Vygotsky, L. S. (1986). *Thought and language.* Cambridge, MA: MIT Press.

PART B

**EMERGENCE OF
SOCIAL OBJECTS AND UNDERSTANDING**

EDITOR'S SECTION INTRODUCTION

Wolff-Michael Roth

From an cultural-historical activity theoretic as from a phenomenological perspective, who we are generally and how we think more specifically cannot be understood independent of the collectivity (i.e., society, culture) that individuals co-constitute and its characteristic practices. But, of course, the collectivity does not exist independent of its individual members. From both cultural-historical activity theoretic and phenomenological perspectives, the relationship between individual and collective has to be thought as mutually constitutive or as singular plural. On the one hand, the collective exists only in and as of its individual members, who constantly realize the existence and conditions of the collective. On the other hand, the collective is the very source of the societal relations that subsequently constitute the higher-order cognitive functions of the individual (Vygotsky, 1978). Society is in the mind as much as mind is in society. The two expressions are but opposite sides of the same coin. We can also use the concept of the *singular plural* for thinking the relation between individual and collective (Nancy, 2000). The collective is a singular (as the singular definite article "the" marks); but it is a plural as well, consisting of all the individuals and their relations. On the other hand, the individual member of the collective is a singular, which, as such, is defined and exists only in and through its relations to all the others. The individual therefore is plural as well, because its membership status is available only in and through the properties common with others. Thus both collective and individual are singular plural entities.

In communities (of practice), as the etymological origins of the term attest (Lat. *com-*, together + *–mūnis*, bound, under obligation), what people do is not independent and singular but literally is what people have in common. Again as the etymology exhibits, being a member of a community means not only having something in common but also means being bound (by) and being under an obligation. This obligation includes, among others, obligations to respect and to intelligibility. This, as I understand the development of mathematics over the past several decades, has been at the heart of the research program that Paul Cobb pursues since the 1980s. He has continuously pushed to reveal (a) not only how students construct/appropriate/make sense of mathematical entities (b) but also how they reproduce and transform the accounting practices, the rules, and the respect for obligations to intelligibility. All three chapters in this Part B of the book are concerned with the emergence and praxis of tacit and explicit rules that are but effects of the underlying obligations to intelligibility in the group as a whole.

In chapter 6 ("Supporting Students' Learning About Data Creation"), Paul Cobb and Carrie Tzou analyze how students in one middle-school classroom came to understand the data creation process and the importance of that process to the drawing of conclusions from statistical data. The data for this investigation were collected during two consecutive classroom design experiments that focused on statistical data analysis. The first design experiment occurred in a seventh-grade mathematics classroom and the second, follow-up study was conducted with a subset of the same student group in their eighth-grade activity period. In analyzing their data, Cobb and Tzou focus both on the students' developing understandings of data creation and on the means by which the emergence of these understandings was supported. That is, I understand the authors to be engaged in a cultural-historical and epistemological project that documents the evolution of both knowledge and accounting practices. In the analyses, the authors show that students did not anticipate ways in which the data creation process might affect their analysis at the beginning of the seventh-grade design experiment. However, towards the end of the eighth-grade experiment, they seemed to have developed a relatively deep understanding of this relationship and played a far more active role in contributing to the data creation process.

In "The Emergence of a Complex Graphical Inscription: The Case of a Bifurcation Diagram" (chapter 7), Chris Rasmussen, Michelle Zandieh, and Megan Wawro take as their starting point the fact of the pervasiveness of interpretations and use of graphical inscriptions in mathematics. Many past studies have examined the difficulties that learners have in interpreting graphical inscriptions and their reluctance to use graphs to solve problems. Whereas these studies add useful insight into student thinking regarding how students' interpret and use graphs, it leaves open the question of how these graphical inscriptions have become mathematical entities for learners in the first place. In this chapter the authors analyze the emergence (or birth) of one such complex graphical inscription known as a bifurcation diagram in a college course on differential equations. In particular, Rasmussen et al. analyze the emergence of this graphical inscription from mathematical, cognitive, and sociological points of view. Important in this process of emergence are brokers and the process of brokering that they enable. Mathematically, the authors examine how learners create a bifurcation diagram. They then investigate the bifurcation creation process in terms of conceptual blending (a cognitive analysis) and in terms of the interactive constitution of meaning (a sociological analysis). The chapter concludes with implications for instructional design and teaching especially in view of their proposal of the idea of brokering.

In chapter 8 ("Inscription, Narration and Diagrammatic Argumentation. Narrative Accounting Practices in Primary Mathematics Classes"), Götz Krummheuer begins with the contention that the global psychological approaches about learning do not describe well the mathematics teaching learning processes in everyday classroom situations since they do not take into account the complexity of such settings. For example, the communication about mathematics is usually bound to specific patterns of a multi-party interaction that is enriched by an inscriptional dimension of mathematics notes, formulas, geometric drawings, and designs. In addition these processes of interaction are specifically shaped by the argumentative demands of the science of mathematics and by the demands on the intelligibility of what speakers say. Based on video-observations of mathematics classes in primary education, Krummheuer reconstructs these complex situations from the perspective of the emergence of the mathematical theme and its co-emerging inscriptional practice and argumentative accounting practice. The author shows that these classroom processes are narratively structured and, in the sense of C. S. Peirce, characterized by diagrammatically based argumentation.

REFERENCES

Nancy, J.-L. (2000). *Being singular plural*. Stanford, CA: Stanford University Press.
Vygotsky, L. S. (1978). *Mind in society*. Cambridge, MA: Harvard University Press.

CHAPTER 6

SUPPORTING STUDENTS' LEARNING ABOUT DATA CREATION

Paul Cobb and Carrie Tzou

Our purpose in this chapter is to analyze how students in one middle school classroom came to understand the data creation process and the importance of that process to the drawing of conclusions from statistical data. The data for this investigation were collected during two consecutive classroom design experiments that focused on statistical data analysis. The first design experiment occurred in a seventh grade mathematics classroom and the second, follow-up study was conducted with a smaller contingent of the same students in their eighth grade activity period. The analysis we present focuses both on the students' developing understandings of data creation and on the means by which the emergence of these understandings was supported. As we document, the students did not anticipate ways in which the data creation process might affect their analysis at the beginning of the seventh grade design experiment. However, towards the end of the eighth-grade experiment, they seemed to have developed a relatively deep understanding of this relationship and played a far more active role in contributing to the data creation process.

Mathematical Representation at the Interface of Body and Culture, pp. 135–170
Copyright © 2009 by Information Age Publishing
All rights of reproduction in any form reserved.

As we have indicated, the two design experiments on which this investigation is based focused on statistical data analysis. The importance of statistics as a core area of emphasis in mathematics and science is indicated by recent reform documents such as the *Principles and Standards for School Mathematics* (NCTM, 2000) and the *National Science Education Standards* (NRC, 1996). The *Principles and Standards for School Mathematics* includes a detailed discussion of data analysis, statistics, and probability as central aspects of curricula for grades six through eight. The authors of these Standards argue that exploratory data analysis should be a precursor to formal statistical inference. In addition, the Standards also emphasize the importance of students' developing inferences and arguments that are based on patterns in the data. Similarly, the importance of statistics to the teaching of science is highlighted in the *National Science Education Standards*, which outline a vision of science for all students. These Standards focus on the teaching of science as inquiry. In particular, inquiry is defined as

> posing questions; examining books and other sources of information to see what is already known; planning investigations; reviewing what is already known in light of experimental evidence; using tools to gather, analyze, and interpret data; proposing answers, explanations, and predictions; and communicating the results. (p. 26)

This characterization frames the process of data collection and data analysis as a central aspect of scientific inquiry. Similarly, the Standards cast the processes of interpreting data and communicating of the conclusions drawn from data as activities that are central to the learning of science.

THE CONTENT OF STATISTICS: EXPLORATORY DATA ANALYSIS, STATISTICAL INFERENCE, AND DATA CREATION

George Cobb and David Moore (1997), in their discussion of statistics teaching, provide a useful vantage point from which to realize the visions outlined in the reform documents described above. The approach taken in the two classroom design experiments is consistent with Cobb and Moore's argument that data analysis comprises three main components: Data creation, exploratory data analysis (EDA), and statistical inference. Although Cobb and Moore are primarily concerned with the teaching of statistics at the college level, the major aspects of their argument also apply to the middle and high school levels.

Exploratory Data Analysis (EDA)

EDA involves the exploration of the batch of data at hand. Cobb and Moore argue that this should be the initial focus of statistics instruction as it is concerned with trends and patterns in data sets and does not involve sample-population relations. Students, therefore, do not need to support their conclusions with probabilistic statements of confidence. Instead, conclusions are informal and based on meaningful patterns that are identified in data sets. As a consequence, when students engage in EDA, they do not have to consider explicitly whether these trends and patterns hold for a larger population.

Statistical Inference

Cobb and Moore argue that EDA is a necessary precursor to statistical inference. Statistical inference involves viewing a data set as a sample and drawing conclusions about the larger population from it. Statements of statistical inference are probabilistic in nature and involve an understanding of sampling distributions, confidence intervals, and significance tests. The intent is to gauge the likelihood that patterns identified in a sample are not specific to that data set but reflect trends in the larger population from which the data were generated.

Data Creation

In teaching statistics, Cobb and Moore (1997) argue that the initial emphasis should be on EDA. Only after students have become familiar with data analysis can they begin to appreciate the relevance of a careful designed data creation process. They observe that "If you teach design before data analysis, it is harder for students to understand why design matters" (p. 816). However, although they recommend introducing design after EDA, Cobb and Moore also recognize that the crucial understandings that students should develop center on the relationship between the legitimacy of conclusions drawn from the data and the soundness of the process by which the data were generated.

Summary

As we clarify below, the approach taken to statistical data analysis in the two design experiments was broadly compatible with the orientation that

Cobb and Moore (1997) outline. Although the primary focus of the two design experiments was on EDA and data creation, issues that could lead to the more formal aspects of statistical inference became, for the research team, an explicit focus of investigation midway through the eighth-grade experiment. For the purposes of this chapter, we do not discuss statistical inference but instead limit our focus to the students' developing understandings of the relationship between data creation and data analysis. In the remainder of the paper, we first give an overview of the two design experiments by discussing the instructional goals, the classroom activity structure, and the nature of the instructional activities. Against this background, we then document how the negotiation of classroom social norms, sociomathematical norms, and the students' role as data analysts served to support their developing understandings of the relationship between data creation and data analysis.

BACKGROUND TO THE ANALYSIS

This analysis is based on data generated during two classroom design experiments, one of 12 weeks and a second, follow-up experiment of 14 weeks. During the fall of 1997, 29 seventh-grade students participated in the first experiment in which a member of the research team served as the teacher. This experiment was conducted in the students' regular mathematics classroom and focused on the analysis of univariate data. In the fall of 1998, a smaller contingent of students from the same class (now eighth graders) participated in a fourteen-week experiment that focused on the analysis of bivariate data with an emphasis on statistical covariation. Between the students' seventh- and eighth-grade years, the school district had implemented a core curriculum that made it difficulty to conduct the second design experiment in the students' regular mathematics classroom. The follow-up design experiment was therefore conducted during the students' activity period and a member of the research team again served as the teacher. Of the original 29 students, eight had transferred to other schools and four had other obligations (e.g., practice for the school play or for the school band). Of the remaining seventeen students, sixteen volunteered to give up their activity period and eleven continued to participate throughout the fourteen weeks of the experiment. Seven of the 11 students who participated for the entire experiment were African American, three were Caucasian, and one was Asian American. An analysis of interviews conducted with all of the 29 original students at the end of the seventh-grade experiment indicated that these 11 students were reasonably representative of the entire group in terms of the diverse ways in which they reasoned about data.

GOALS OF THE CLASSROOM DESIGN EXPERIMENTS

The overall goal of the two design experiments was to support the emergence of increasingly sophisticated ways of analyzing univariate and bivariate data sets as part of the process of developing effective data-based arguments. The overarching mathematical idea that served to orient instructional design in both experiments was that of distribution. The potential endpoint for the seventh grade design experiment was that the students would come to describe how univariate data were distributed within a space of values. Notions such as center, skewness, spread-outness, and relative frequency then would emerge as ways of describing how specific data sets are distributed within this space of values. In the eighth-grade design experiment, the research team planned to build on the students' prior experience of analyzing univariate data sets so that they might come to view bivariate data sets as distributions within a two-dimensional space of values. Notions such as the direction and strength of the relationship between the two sets of measures made when generating the data would then emerge as ways of describing how the data were distributed within this space of values. The research team was relatively successful in achieving these goals.

TALKING THROUGH THE DATA CREATION PROCESS

The classroom activity structure for both the seventh- and the eighth-grade design experiments consisted of three main phases. The first phase was a discussion of the process of data creation in which the teacher and students talked through how they could generate data that would enable them to address a particular problem or issue. When they formulated this approach, the members of the research team viewed these discussions of the data generation process as essential as the students would not, for the most part, collect data themselves. In the second phase of the classroom activity structure, the students worked either individually or in small groups to analyze data sets that the teacher introduced as resulting from this data generation process. In conducting these analyses, the students used one of a series of three computer-based analysis tools that were introduced sequentially during the two design experiments. The third and final phase of the activity structure was a whole-class discussion in which the students used a computer projection system to make their analyses public for critique and discussion. Because the first phase of the classroom activity structure, the data creation discussion, is particularly important for our analysis we describe its rationale in some detail.

The research team's primary goal while conducting the two design experiments was to support the students' development of increasingly sophisticated ways of organizing and structuring data (i.e., data analysis). In light of this focus and the time constraints of both design experiments, the research team made the conscious decision that the students would not, for the most part, collect the data they were to analyze. However, the team was also aware of the danger that if the students did not collect data, the data values might be mere numbers for them rather than measures of attributes of a phenomenon. A pilot study conducted prior to the start of the seventh grade design experiment had revealed that students frequently compared data sets by calculating the means in order to reduce a set of numbers to a single number. In doing so, they seldom reflected on either the meaning of the result of this calculational procedure or the appropriateness of this means of comparing data sets to the question at. It appeared that the data values were numbers to be manipulated for them rather than measures of an attribute of a phenomenon that was relevant to the question at hand. In light of this finding, the research team conjectured that it would be essential for the teacher and students to talk through the data creation process so that the data sets the students were to analyze had a history that reflected the interests and purpose for which they were generated. From the research team's perspective, it was essential that the students come to view data as sets of measures that could be analyzed to address particular questions and issues rather than as sets of numbers to be manipulated because they were asked to do so by the teacher.

The emphasis that the research team placed on the data generation process reflects the view that data do not simply come ready-made. Instead, data are typically to address specific problems or issues and reflect the choices that investigators make when they construe a situation. In addition, as Gravemeijer (1999) notes, data are usually generated with a certain audience in mind. The issues addressed during discussions of the data generation process therefore included the purpose for generating the data, the overarching problem or issue being addressed, and who would benefit from addressing this problem. The teacher typically initiated these discussions by introducing a general problem or issue. In the ensuing conversation, the teacher and students clarified why this problem or issue would be significant to them or to a particular audience. Next, the students and teacher discussed which attributes of the phenomenon under consideration were relevant to the issue they sought to address. Frequently, they also discussed how they would actually measure these attributes in order to generate the required data. It was not until these issues had been resolved that the teacher introduced the data the students were to analyze. In doing so, she often discussed the significance of possible data values with respect to the situation from which the data were generated (e.g., the significance of a resting heart rate of 90 beats per minute in an investigation of the effectiveness of exercise programs for patients' heart problems). The intent of these discussions was that the students would become familiar with how, why, and for whom the data were generated so that they might be able to interpret the variability of data values in terms of the situation from which the data were generated.

This approach of talking through the data creation process was inspired in part by Roth's (1996) study in which he found that students were unable to analyze data when they had not been actively involved in its generation. The students who participated in this study had designed their own research projects during a ten-week ecology unit. Against this background, Roth asked them to solve a story problem that involved a situation with which they were already familiar from their own research projects in the ecology unit. The students were given a map of an ecozone on which were recorded the same types of data that they were accustomed to collecting in their own investigations. Roth reports that the students were unable to analyze the data presented in this manner because they did not have direct access to the situation from which the data had been generated, as was the case when they conducted research projects.

The research team found Roth's results particularly significant given that the students in the design experiments would analyze data that had already been inscribed in one of the three computer minitools as case magnitude plots, line plots, or scatter plots, respectively. The students might therefore find themselves in a situation similar to that experienced by Roth's students when they were presented with data that had already been inscribed on the map of the ecozone. Roth's analysis therefore oriented the research team to focus explicitly on both the process of generating data by measuring aspects of a phenomenon and the process of transforming data by inscribing them in various ways. As Roth (1996), observes, this process produces a cascade of inscriptions that progressively objectify the data by distancing them from the original situation. He gives the following illustration to clarify this process of inscribing "raw data" (i.e., measures) in a manner that makes them amenable to analysis:

> To work on and think about natural phenomenon (i.e., to mathematize phenomena), scientists engage in practices that ultimately produce inscriptions. Latour (1993) showed how a phenomenon and its mathematical order were produced as members of a scientific expedition first took soil samples that were subsequently placed in a two-dimensional array of boxes according to location and depth of the probe. This practice, one among many in which the members of the expedition engaged, made the phenomenon accessible to transformation into various drawings on paper (inscriptions) and to analysis wherever the drawings could be transmitted (e.g., labs in different parts of the country). (p. 493)

Roth's (1996) students experienced difficulty when they attempted to solve the story problem because the map was not, for them, the product of a transformative process of this type. In other words, the situation from which the data had been generated was not transparent in the map for the students because they could not take the process that led to is creation as a given. As a consequence, "a meaningful situation was changed into a puzzle with few options" (p. 518). This, for the research team, indicated that it would be important that the students in the design experiments enact the data creation process before they began an analysis. However, the team also conjectured that it would not be crucial for the students to actually collect data themselves as had Roth's students when they conducted their research projects. Instead, the team speculated that it would be sufficient for the teacher and students to talk through the data creation process. If this conjecture proved viable, data inscribed in the computer minitools would be texts of the situations from which the data were generated for the students.

INSTRUCTIONAL ACTIVITIES AND THE COMPUTER TOOLS

Given the emphasis on EDA in the design experiments, the research team considered it essential that the students' activity in the classroom involve the genuine spirit of data analysis from the outset. This implied that the instructional activities should involve the analysis of realistic data sets for purposes that the students would consider reasonable. As a consequence, the research team developed instructional activities that involved either (a) analyzing a single data set in order to understand a phenomenon, or (b) comparing two data sets in order to make a decision or judgment.

As we have noted, the students used three computer-based tools that had been developed by the research team to create and manipulate graphical inscriptions of data. The first minitool, introduced in the fifth classroom session of the seventh-grade design experiment, was designed to provide students with a means of ordering, partitioning, and otherwise organizing sets of up to 40 univariate data points. When data are entered into this minitool, each individual data point is inscribed as a horizontal bar, the length of which signifies the numeral value of the data point. The students could select the color of each bar to be either pink or green, enabling them to enter and compare two data sets. Figure 6.1 shows the data for one of the first instructional activities in the seventh grade design experiment in which the students compared two brands of batteries, "Always Ready" and "Tough Cell," in order to decide which brand was superior.

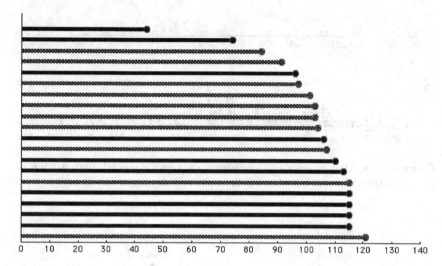

Figure 6.1.   "Tough Cell" and "Always Ready" battery data on the first computer minitool.

The data are measurements of the life span of ten batteries of each brand. In this case, the Always Ready batteries were inscribed as green bars and the Tough Cell batteries were inscribed as pink bars.

The second computer minitool was introduced in the twenty-second session of the seventh-grade design experiment. The tool was designed to allow students to analyze one or two univariate data sets of up to 400 points. Individual data points were inscribed as dots located on a horizontal axis of values. The dots at the end of the bars in the first tool had, in effect, been collapsed down onto the axis. Figure 6.2a shows data from an instructional activity which was used towards the end of the seventh-grade design experiment in which the students were to analyze the T-cell counts of AIDS patients who had enrolled in two different treatment protocols, an experimental treatment and a standard treatment. The data for the two treatments are inscribed as two colors, with the standard treatment inscribed as green dots and the experimental treatment inscribed as pink dots.

In the third computer minitool, bivariate data are inscribed as scatter plots. This tool was introduced in the eighth-grade design experiment after the teacher and students had established that bivariate data consist of the measures of two attributes of each of a number of cases. Figure 6.3 shows the data from the last instructional activity of the eighth-grade design experiment, in which students analyzed pre- and post-test scores

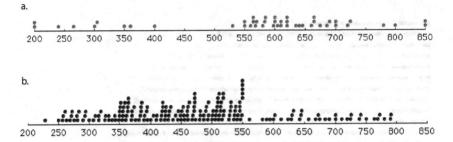

Figure 6.2.   a. AIDS experimental treatment data on the second computer mini-tool. b. AIDS traditional treatment data on the second computer minitool

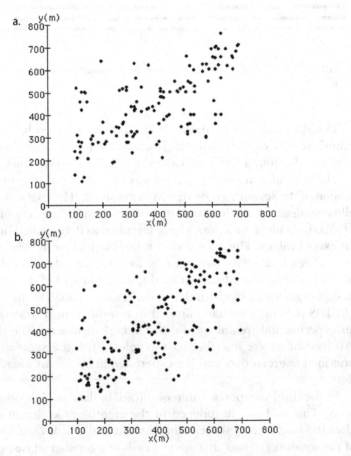

Figure 6.3.   a. Speed-reading data for Program G1 on the third computer mini-tool. b. Data from speed-reading course G2 on the third computer minitool.

for two speed-reading courses. The data are inscribed on the scatter plot as measures of individuals' reading speed in words per minute before (on the X-axis) and after (on the Y-axis) attending the speed-reading course.

DATA SOURCES AND METHOD OF ANALYSIS

All 34 lesson of the seventh-grade experiment were video-recorded, as were all forty-one lessons of the eighth-grade experiment. Two cameras were used: One camera focused on the teacher and students as they came to the front of the classroom to explain their analyses. The second camera focused on the students during both the data creation discussions and the concluding data analysis discussions. The primary data sources for the analysis reported in this chapter consisted of the video-recordings of these discussions, two sets of field notes all the classroom sessions, and video-recorded individual interviews conducted with the students at the end of the eighth-grade design experiment.

The method for analyzing these data involved first identifying critical incidents from the two sets of field notes. The criteria used to identify critical incidents included finding episodes in which either the teacher and students appeared to be negotiating the relation between data generation and data analysis or a shift appeared to have occurred in the way that students contributed to the data generation discussions. Once an episode was identified, the videotape segment was reviewed and transcribed if new understandings of the data generation process and its relation to data analysis appeared to emerge. In this manner, conjectures about the students' understanding of the data generation process that were formulated while analyzing one episode were tested and revised on an ongoing basis while analyzing subsequent episodes. This interplay between conjectures and refutations constituted the primary means for establishing the trustworthiness of the analysis. Further, the general assertions about the evolution of the students' understandings of data generation developed in this way are empirically grounded in the data corpus and can be substantiated by backtracking to the field notes, video segments, and transcripts. This approach for analyzing large corpuses of classroom video-recordings is consistent with an adaptation of the constant comparative method that is tailored to problems and issues in mathematics education (Cobb & Whitenack, 1996).

ANALYSIS

The purpose of this analysis is to account for how the students came to develop a reasonably sophisticated understanding of the relationship between data generation and data analysis. As we mention above, the students came to understand that the legitimacy of the conclusions they could draw from the data depended on the soundness of the data generation process. In the following paragraphs, we present two contrasting classroom episodes, each from the data generation discussions, to illustrate the changes that occurred in the students' reasoning. The first episode comes from the beginning of the seventh-grade design experiment, and the second from near the end of the eighth-grade design experiment. We then summarize the findings of individual interviews conducted at the end of the eighth-grade experiment to substantiate our claims about the students' learning.

In the third session of the seventh-grade design experiment, the teacher introduced an instructional activity that involved braking distances of cars. These data were first presented to the students as part of a homework assignment and then in class inscribed as horizontal bars in the first of the three computer minitools. The data were measures of the braking distances of ten cars from Company A and ten cars from Company B measured in feet (see Figure 6.4). The students were asked to decide which of the two types of cars they would recommend that someone who was particularly concerned about safety should buy. During the data generation discussion, the students' comments were mainly of an anecdotal nature and did not relate to the data generation process. At this early stage of the first design experiment, the teacher and students

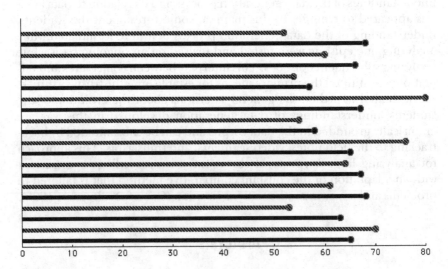

Figure 6.4.   Data for the Braking Distances activity inscribed on the first computer minitool.

appeared to have sharply contrasting understandings of the intent of data generation discussions. The following representative excerpt occurred immediately after the teacher had outlined the general braking distance scenario.

1	Kevin:	They have a commercial where it's like a big long line and they have the tires
2		and says this is what happens when the tire like stops. And showing how fast it takes
3		to go from 60 to 0.
4	Teacher:	Yeah as opposed to 0 to 60. Usually, yeah, usually the cars, that you know
5		the sports cars say from 0 to 60 in you know so many seconds. They don't talk a
6		whole lot about from 60 to 0 which is pretty important when you have to stop.

In lines 1–3, Kevin comments on a television commercial that relates to the topic that the teacher has introduced, but not to the process of generating data. Later in the discussion, the teacher presented the data sets to the students and encouraged them to ask questions or make comments before they began their analyses.

7	George:	When my sister was taking her driving test, I used to watch her videos along
8		with her. And they said on an icy road it could take a car 275 feet to stop.
9	Teacher:	To stop. That's exactly correct. Because it just starts sliding. There's just
10		nothing for the wheels to grab hold of.

Frequent comments of this type indicate that the issue under discussion for the students was that of the braking distances of cars per se rather than the generation of data that would enable them to compare the safety of the two types of cars. From the research team's point of view, the students' contributions were anecdotal in that they did not appear to reflect an awareness that the process by which the data were generated could effect the subsequent analysis. This was, however, to be expected as the students had little experience with analyzing data at this point in the design experiment.

This data generation discussion contrasts sharply with those that occurred during the latter part of the eighth-grade design experiment. Towards the middle of this experiment, the students were involved in a sequence of instructional activities in which they investigated a number of issues relating to driver safety (e.g., reaction times for drivers of different ages, insurance rates for different age groups, and the adequacy of Tennessee's law for driving under the influence of alcohol). Prior to the episode described below, the students had tested their own reaction times by measuring how long it took them to catch a meter ruler that was dropped through an open hand by another student. The teacher introduced an instructional activity that focused on the relationship between alcohol consumption and reaction time by explaining that the data had been generated by giving people certain amounts of alcohol to consume and testing their reaction times in a similar manner' by using a meter stick (see Figure 6.5).

1	T:	Remember we talked about how they collected this data. They had a group of 23
2		year olds and they gave them some liquid to drink and they didn't know how much
3		alcohol was in it. Some of them had 0 ounce, some 1, 2, 3, 4 so on. So they got these

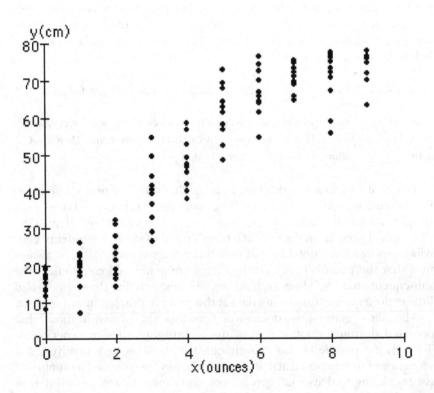

Figure 6.5.   Data for the alcohol and reaction time activity inscribed on the third minitool.

4	different reaction times. And one of you pointed out that was really a good idea
5	because sometimes when you know you're drinking a lot of alcohol, it inadvertently
6	affects the way you act. But they didn't know, so then they got these reaction times on them.
7	Jerod: They didn't know?
8	Teacher: They didn't know how much was in there.

In line 7, Jerod asked a clarifying question pertaining to the way in which the data were created. By asking, "They didn't know?" (line 7), he seemed to indicate his understanding that if the participants had known how much alcohol they were drinking, their performance on the reaction time test could have been effected. The data creation discussion continued:

9	Teacher: I'm sure that they knew if there was none or some, although I don't even
10	know that since I don't know how it was mixed, but they didn't know the amount of
11	alcohol that they had had when they did the test. Yes ma'am?
12	Brenda: Did they know they were drinking alcohol?
13	Teacher: They signed up for a test in which they knew that they could be given an
14	amount of alcohol.
15	Brenda: Are these people who normally drink or are they people who don't drink?
16	Teacher: These are people ... oh that's ... probably some of both, because what they
17	were trying to do was get a sample of people to see what happens when people ... what
18	happens to their reaction times when they drink.

The question Brenda asked (line 15) suggests that she might have been aware of the importance of testing a representative group of people. In particular, she seemed to recognize that if the participants were not normally drinkers, or if they normally drank large amounts of alcohol, their reaction times might not have been representative of the general population. In general, the students took a much more proactive role in this and other data generation discussions and they appeared to have some appreciation of its relevance to the analysis they were to conduct.

The changes in the students' reasoning about data generation that we have illustrated by presenting the two sample episodes were substantiated by video-recorded individual interviews conducted shortly after the end of the eighth-grade design experiment. All eleven students who participated in the entire eighth-grade experiment were interviewed, as were nine students who were not involved in either this experiment or the preceding seventh-grade experiment. As one of our goals was to pilot questions that might later be used in paper-and-pencil assessment instruments, the interviews were based on one of four different data generation scenarios. For ease of explication, we summarize the findings by referring to only one of these scenarios:

> The coach of a men's high school football team wants to know how the number of calories players eat the day of a game affects their energy during the game. He'd like to know how many calories his players should eat on game day to feel their most energetic. (The average adult male consumes about 2,500 calories a day. Adults trying to lose weight often cut back to about 1,000 calories a day).

The interviewer asked the students what data they would collect if they had been asked to investigate this question. Eight of the 11 design experiment students described the types of quantifiable data they would collect (e.g., the number of calories consumed by a group of players and the number of points scored). One of the remaining three students proposed a type of experimental design in which the players consumed different amounts of calories in different games. A second student seemed to interpret the question as that of finding the effect of different types of food and proposed having the players record both what they ate and their subsequent "mood." The final student interpreted the question as that of comparing two different types of food and focused on the players' performance but did not mention measuring the number of calories consumed. In answering this question, several of the students also mentioned that it would be difficult to measure a player's energy level and one proposed using a self-rating scale of one to ten. Some of the students also talked about other factors that might affect energy level and that should be controlled in some way. This led one student to propose a type of experimental design in which groups of players would consume differing amounts of calories but without knowing how many calories they had consumed.

The students were next asked how they would organize the data they proposed to collect. Seven of the eight students who had proposed to measure calories consumed and performance during a game explained that they would use a scatter plot. Follow up questions clarified that a dot on the scatter plot would signify a player and that the variables would be calories consumed and performance. The remaining student who also proposed to collect data of this type said that she would make a chart and indicated that she considered it important to personalize the information by organizing it by individual player. The three design experiment students who said that they would collect different types of data were relatively vague about how they would organize the data.

The interviews conducted with the 11 teaching experiment students were all based on scenarios that, potentially, involved the generation of bivariate data. In contrast, the interviews conducted with six of the nine students who did not participate in the two design experiments were based on a scenario that, potentially, involved the generation of univariate data. The issue they were asked to address in this scenario was that of determining the relative effectiveness of two brands of headache medication in giving relief to migraine sufferers. Only two of these nine students proposed to collect quantifiable data of some type and both were questioned about the migraine scenario. One of these students proposed a design in which a group of volunteers would take one drug the first time they had a headache and the second drug on the next occasion that they had a headache and would record the time until they felt relief. Although this student spoke of organizing the data by drawing a bar graph, he was not able to clarify how he would use it or why it might be useful. The other student proposed to collect quantifiable data also proposed to measure time to relief, but intended to test each drug on only one person. The remaining seven students tended to raise issues that were, from the interviewer's perspective, extraneous to the question at hand. Further, follow-up probes indicated that concerns such as the selection of a sample or procedures for measuring attributes of a phenomenon were not issues for them.

As a general summary, most of the design experiment students were able to engage in a data generation conversation in which they addressed issues such as the adequacy of the experimental design, the measures used, possible confounding variables, ways of organizing the data, and conjectures about what the data might show. Only two of the students who did not participate in the experiment seemed to be able to engage in a conversation of this type, and then only when considering the experimental design and the measures used. The contrast between the two groups of students was so marked that we in fact concluded we could not pose questions of this type on paper-and-pencil instruments. Because the students who had not participated in the design experiments could not, for the most part, even begin to engage in a data generation conversation, the questions served to reveal only their limitations.

In the remainder of this chapter, we account for the changes we have documented in the design experiment students' reasoning about data generation and in the ways in which they participated in data generation discussions. In doing so, we consider three means by which their emerging understandings of the relation between data creation and data analysis were supported: (a) The social norms that emerged in the design experiment classroom, (b) the sociomathematical norm of what counted as a data-based argument, and (c) the students' emerging role as data analysts. We have listed these three aspects in chronological order as the emergence of one aspect depended on the emergence of prior aspects. For example, the emergence of students' role as data analysts was made possible by the prior establishment of both classroom social norms and sociomathematical norms.

CLASSROOM SOCIAL NORMS

The development of the students' understanding of the relationship between data creation and data analysis can be attributed in part to the social norms that were established during the concluding data analysis discussions in the first few sessions of the seventh-grade design experiment. During these discussions, the teacher made it clear that she expected the students to understand each other's analyses and to ask clarifying questions when they did not understand. The teacher conveyed this expectation to the students by assuming that if they did not ask questions they would be able to explain each other's analyses and reasoning.

As an illustration, consider the following episode that occurred during the concluding data analysis discussion of the braking distance activity (see Figure 6.4). Early in this discussion, one student, Melissa, began to explain her analysis of the data to the class. However, it was soon apparent that most of the students did not understand her explanation. At this juncture, another student, Kevin, asked if he could restate Melissa's explanation.

1	Kevin:	Ok, I'm just going to restate what Melissa said.
2	Melissa:	Thank you, I'm having problems
3	Teacher:	Ok, this will be helpful.
4	Kevin:	I know …
5	Teacher:	Wait, wait. Can I clarify? Melissa, your job is to make sure that you agree
6		with what he's saying that you're saying. Ok, that's your job. And everybody else is to
7		listen and see if you understand.

The teacher, in lines 5 through 7, made the students' obligation to actively listen and attempt to understand each others' explanations explicit. As this exchange indicates, the teacher encouraged the students to restate each other's explanations both to help others understand and to check their own interpretations. The teacher further emphasized these obligations of active listening and striving to understand each other's explanations as the discussion progressed. In the following excerpt, one student, Drew, had just explained his analysis in which he noticed that the braking distances of the cars of one brand were more consistent than the braking distances of the cars of other brand.

1	Teacher:	Are there questions, guys, are there questions for Drew. Ok. Nobody has a
2		question. That means that I could call on anybody in here and you could explain what
3		Drew just said. Troy. Troy has the floor.
4	Troy:	He said that the pink one, he said that's more consistent.
5	Student 1:	What does consistent mean?
6	Teacher:	Ok. What does consistent mean?
7	Drew:	It means it ...
8	Student 2:	It gives the same thing every time or close to it.
9	Teacher:	Ok. Jerod, you have something to add? Ok. So that's a really nice argument
10		that Drew is making that it's... he thinks that if you own a car, it's kind of important
11		to know that every time you stop.... Every time you stop it's going to stop within
12		range.

In lines 1 and 2, the teacher again made her expectations for the students explicit when she said, "Nobody has a question. That means that I could call on anybody in here and you could explain what Drew just said." As a student's request for a clarification of the term "consistent" (line 5) indicates, the teacher was relatively successful in initiating the renegotiation of classroom social norms early in the seventh-grade design experiment.

The emergence of these social norms for the data analysis discussions is relevant to the purposes of this investigation in that it also appeared to legitimize the asking of clarifying questions during the data generation discussions. However, the emergence of these norms did not orient students to the specific aspects of data creation that would be important to interrogate. Thus, although the social norms for data analysis discussions gave the students license to question each other's contributions to the data creation discussions, they provided the students with little guidance as to how they might do so in ways that related to the teacher's agenda. At this point in the experiment, the data creation discussions typically consisted of a series of discrete, separated turns because the students rarely referred to each other's largely anecdotal contributions.

SOCIOMATHEMATICAL NORMS

Social norms such as those that we have discussed for the data creation and data analysis discussions constitute what Erickson (1986) referred to as the classroom participation structure. Sociomathematical norms differ from general social norms that constitute the classroom participation structure in that they concern the normative aspects of classroom actions and interactions that are specifically mathematical. Examples of sociomathematical norms include what counts as a different mathematical solution, a sophisticated mathematical solution, and an efficient mathematical solution. These three norms all relate to students' understandings of when it is appropriate to contribute to a discussion. In contrast, a fourth sociomathematical norm that is particularly important for our analysis, that of what counts as an acceptable mathematical argument, deals with the actual process of making a contribution.

The teacher and students explicitly negotiated the sociomathematical norm of what counts as an acceptable data-based argument relatively early in the seventh-grade design experiment. In the first data analysis discussion involving the first computer minitool, the students spoke almost exclusively of "pinks" and "greens" when they referred to the data inscribed as horizontal bars in the minitool. Further, they did not explain why the ways in which they had structured the data were relevant with respect to the question at hand. It therefore seemed that an appreciable number of the students were describing differences in two sets of numbers inscribed as colored bars rather than analyzing data. However, in the second instructional activity involving the minitool, the teacher was able to initiate a shift in the classroom discourse such that the students began to justify why the ways in which they structured the data were relevant to the question that they were addressing. In this instructional activity, the students analyzed data on the life spans of ten batteries of each of two brands, "Always Ready" and "Tough Cell," in order to decide which brand of battery they would recommend (see Figure 6.1). The first student to explain her reasoning in the data analysis discussion indicated that she had focused on the ten highest values:

1	Caitlin:	And I was saying see like there's seven green that last longer ...
2	Teacher:	Ok, now the greens are the Always Ready, so let's make sure we keep up
3		with which set is which, ok?
4	Caitlin:	Ok, the Always Ready is more consistent with the seven right there, and then
5		like seven of the Tough ones, they're like ... I was just saying cuz like all seven, seven
6		out of ten of the greens were the longest, and there was ...

7   Jane: Good point. I understand.

8   Teacher: You understand? Ok, Jane, I'm not sure I do. So could you say it for me?

9   Caitlin: She's saying that out of ten of the batteries that lasted the longest, seven of

10   them are green, and that's the most numbers, so the Always Ready batteries are

11   better, because most of those batteries lasted the longest.

In this episode, Caitlin began by talking about "the greens." The teacher interjected immediately by re-voicing "the greens" as "the Always Ready" in an attempt to ensure that the students remained grounded in the situation of comparing the two brands of batteries. Jane acted in accord with the teacher's expectation when she restated Caitlin's explanation by clearly speaking of batteries as she interpreted the graph (see Figure 6.6). As the episode continued, another student questioned Caitlin's analysis by saying that if she had looked at the highest fourteen rather than just the highest ten values, the two brands of batteries would have looked more comparable. The teacher then asked Caitlin:

12   Teacher: Well maybe Caitlin you can explain to us why you chose ten. That would be

13   really helpful.

14   Caitlin: Alright, there was ten of the Always Ready, and there was ten of the Tough.

15   So that's 20, and half of 20 is 10, so that's how I chose it.

16   Teacher: Why would it be helpful for us to know about the top 10? Why did you

17   choose that? Why did you choose 10 instead of 12?

18   Caitlin: Because I was trying to go with the half.

By asking Caitlin why she chose "10 instead of 12," the teacher attempted to indicate that data-based arguments should include a justification that related the way of structuring the data to the issue under investigation, in this case that of deciding which of the two brands of batteries was superior. It is clear from Caitlin's response, however, that she could not justify her focus on the ten highest values in these terms. In contrast, later in the discussion, another student offered an explanation that did satisfy the teacher's expectation:

19   Casey: Now see there's still green ones behind 80, but all of the Tough Cell is above

20   80. So I'd rather have a consistent battery that I know will get me over 80 hours than

21   one that just try to guess.

22 Teacher: Why were you thinking 80?

23   Casey: Well because most of Tough Cell batteries were all over 80

24 Teacher: Ah. Ok. So it's like a lower limit for you. Questions for Casey? Yes Jane?

25   Jane: Um, why wouldn't the Always Ready batteries be consistent?

26   Casey: Well because all your Tough Cell is above 80, but you still have 2 that are

27   behind 80 in the Always Ready.

Three aspects of this explanation are relevant to our analysis. First, although Casey spoke of "green ones" once in his explanation, it seems clear that he is comparing the life spans of the batteries in his explanation

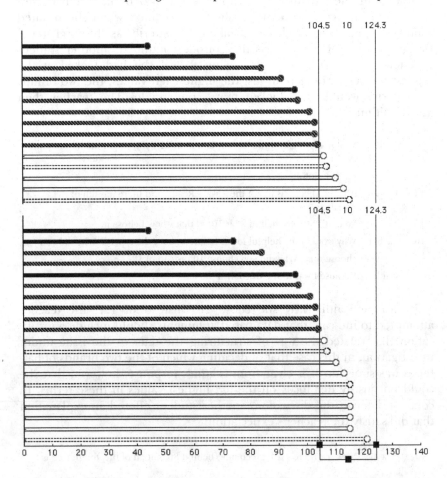

Figure 6.6.   Caitlin's way of structuring the battery data on the first computer minitool.

rather than pink and green bars. A second relevant aspect is the way that Casey related his explanation to the issue of which battery he would choose. He interpreted a feature in the inscription—80 hours—as indicating a difference in the two brands of batteries that he considered significant (see Figure 6.7). Additionally, he gave the rationale for his argument ("I'd rather have a consistent battery that I know will get me over 80 hours than one that you just try to guess") without being explicitly asked to do so by the teacher. A third significant aspect of this explanation occurred in line 25 when Jane challenged Casey's argument, and he again justified his argument in terms of the way he structured the data.

This exchange is representative of the process by which the sociomathematical norm of what counts as an appropriate data-based argument emerged. Because the social norm of asking clarifying questions had already been established for the data analysis discussions, the teacher was able to question the students about their reasons for structuring the data in particular ways. When the teacher asked Caitlin, "Well maybe Caitlin you can explain to us why you chose ten" (line 12), she was asking a clarifying question, one that was in accord with the social norms that had been established in this classroom. In addition, by asking such questions, the teacher guided the renegotiation of the sociomathematical norm of what counted as an acceptable data-based argument. The emerging norm involved justifying the way in which the data had been structured so that it was relevant to the issue at hand. This norm remained relatively stable

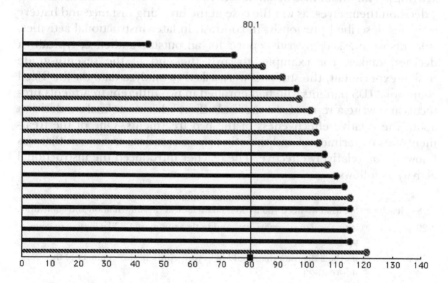

Figure 6.7.   Casey's way of structuring the data for the battery task on the first computer minitool.

throughout both the remainder of the seventh-grade experiment and throughout the eighth-grade experiment.

This development was significant with respect to the students' understandings of the relation between data creation and data analysis in that they were obliged to justify the way they structured data during analysis with respect to the purposes for which the data had been created. The establishment of this sociomathematical norm constituted an initial link between data creation and data analysis. However, this relation was retrospective in that the students were obliged to take account of the prior process of data creation when they conducted their analyses. They were yet to anticipate how the data creation process might influence the conclusions that could be legitimately drawn from the data.

THE STUDENTS' ROLE AS DATA ANALYSTS

Midway through the seventh-grade design experiment, the research team explicitly attempted to support the emergence of the students' role as data analysts. The data analyst role required that the students do more than analyze the data in order to make a decision or judgment themselves. They also had to report relevant trends that they identified in the data in a readily understandable manner to decision makers who were to act on the basis of their analyses. To support the emergence of the students' role as data analysts, the research team modified the instructional activities. The initial instructional activities involved the students making a decision themselves, as was the case in the breaking distance and battery activities described previously. In contrast, in later instructional activities, students were asked to write reports for an outside audience, typically a decision maker. For example, towards the end of the seventh-grade design experiment, the students were asked to analyze data on the T-cell counts of AIDS patients who had enrolled in two different treatments protocols and write a report for a medical officer who was to make a decision about the relative effectiveness of the two treatments. One of the treatments was experimental and the other was a standard treatment that was known to be relatively effective. The teacher introduced the instructional activity as follows:

28	Teacher:	So, what we need you to do is take a look at this information and come up
29		with a way to summarize it. And in that summary I'd like for you to come up with
30		maybe a diagram that would help you represent the summary of the information
31		you could then give to someone else that could then make the decision. And when

32	you're doing this it's real important to remember that that person is not going to have
33	these graphs. All they're going to have is the information that you give them.
34	whatever you put on your diagram and whatever you put in your paragraph will be
35	the only thing that they can use to make the decision.

The teacher therefore expected the students to both analyze the data by identifying patterns in the data that were relevant to the question under investigation and to report them in a manner that would be comprehensible to a person who had not seen the data. Consequently, it was essential that the students, in their role as data analysts, understood both to whom they were writing their reports and the purpose for which their reports would be used.

The extent to which the students subsequently assumed the role of data analysts is indicated by the fact that they subsequently refused to initiate analyses unless they completely understood both the situation from which the data were generated and the issue that the decision maker needed to address. The first such instance occurred at the near the end of the seventh grade design experiment when the students were asked to analyze data on the average daily high temperatures in four cities: San Francisco, Anchorage, Phoenix, and Nashville (see Figure 6.8). As the research team had intended that this activity would serve as a performance assessment, the teacher did not discuss the purposes for which the students' analyses would be used. The data were presented as line plots, an inscription with which the students were familiar, and they were asked to write a report describing trends in the temperatures of the four cities.

36	Kevin:	Here's what I see is just the highs, it's like ... there's no way to ... this makes no
37		sense. How about that? It just makes no sense.
38	Teacher:	Is that a statement or a question?
39	Kevin:	It's a statement.
40	Teacher:	Ok, well you need to come up with a question so that I can help you
41		understand it Kevin. What part makes no sense?
42	Kevin:	Just the whole idea. It doesn't even tell us what we're supposed to do.
43	Student:	It says write a report
44	Katie:	Oh, I know what he's saying
45	Teacher:	What Katie?
46	Katie:	He's trying to say what is the problem. Why?
47	Teacher:	Why would anybody want to know this?

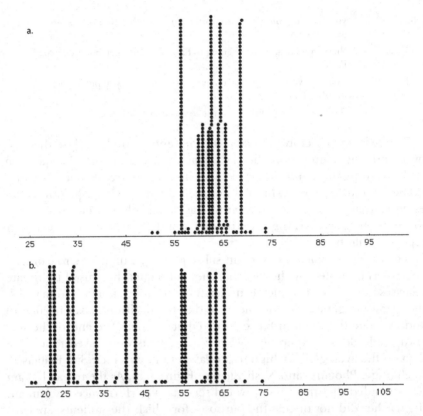

Figure 6.8.   a. Temperature data for San Francisco, California. b. Temperature data for Anchorage, Alaska.

In saying "It just makes no sense" (lines 36–37), Kevin expressed his inability to see the reason for analyzing the data. Kevin's comment is representative of the views that other students expressed in the remainder of the discussion and indicates that, from their point of view, this instructional activity violated an established norm of what it meant to analyze data. This constitutes relatively strong evidence that the students, in their role as data analysts, needed to understand the purpose for which their analyses were to be used.

A second example that illustrates the students' need to understand the situation from which the data were generated occurred in the middle of the eighth-grade design experiment when they were asked to analyze data on car insurance rates for different age groups of people. The students' task was to investigate if data on accident rates justified higher insurance rates for younger drivers than for older drivers. However, they were

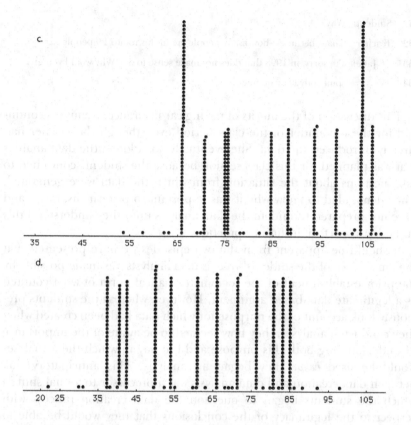

Figure 6.8.   c: Temperature data for Phoenix, Arizona. d: Temperature data for Nashville, Tennessee.

unfamiliar with the practices of insurance and, as a consequence, did not understand the probabilistic notions that underpin car insurance. They therefore questioned why anybody would pay for car insurance in the first place and why an insurance company would pay a claim that is greater than the total premiums an individual has paid. This discussion was initiated when a student asked: "I'm not going to be able to understand this because I have no idea anything about insurance. So can somebody please explain what a policy premium is to me?" A few minutes later, the following exchange occurred:

48   Jane: I'm confused. So basically you just give the company [money] in case you
49   an accident and if you do then they give it to you.
50   Teacher: Not your money. They can give you far more than you've paid them. So …

51   Student: Why?

52   Teacher: That's because—how many people are in this room? 20 people ...

53   Jane: I'm sorry, in 1998 that does not make sense to me. Why would you give

54   somebody your money ...

This discussion of the merits of paying car insurance premiums contin-ued for the remainder of the class period even though the teacher had already introduced the data. She eventually abandoned the data analysis that was planned for this class session because the students continued to ask questions about the situation from which the data were generated. They demanded to know why it was important to pay car insurance and were not prepared to initiate the data analysis until they understood this and other aspects of insurance practices.

It should be apparent from the two episodes we have presented that the emergence of the students' role as data analysts was made possible by the prior establishment of the sociomathematical norm of what counted as a legitimate data-based argument. However, whereas the students pre-viously took account of the purpose for which data had been created when they conducted analyses, they now seemed to be aware of the importance of understanding both this purpose and the way in which their analyses would be used *before* they began an analysis. This anticipatory link between data creation and data analysis was a precursor to a final shift in which the students began to question the data creation process with respect to the legitimacy of the conclusions that they would be able to draw from the data.

DEPENDENCY OF CONCLUSIONS ON DATA CREATION

As we document here, it was essential at the beginning of the seventh grade design experiment for the teacher to play a central role in guiding discussions of the data creation process. The teacher introduced the situation from which the data were generated and was responsible for, in a sense, telling a story that recreated the data creation process. We also doc-ument that the students' initial contributions were largely anecdotal and that they and the teacher had differing understandings of the intent of data creation discussions. However, as the role of data analyst emerged in the latter part of the seventh-grade design experiment, the students insisted on understanding both the situation from which the data had been generated and the way in which their analyses would be used. These developments involved a gradual hand-over of responsibility from the teacher to the students during the data creation discussions.

As the eighth-grade design experiment progressed, the students began to anticipate possible limitations of the data creation process that might affect the legitimacy of conclusions they could draw from the data. For example, the students became increasingly concerned with controlling extraneous variables when they contributed to data creation discussions. In addition, they raised questions that related to issues of sampling and the representativeness of the data. In voicing these concerns, the students contributed to the development of the data creation process as a joint narrative. This contrasts sharply with the data creation discussions at the beginning of the seventh-grade design experiment in which the students neither asked clarifying questions nor attempted to ascertain the purpose of the analysis.

We previously illustrated the types of questions that the students asked at during the latter part of the eighth grade experiment when we presented an episode from an instructional activity in which they analyzed data on the reaction times of individuals who had consumed varying amounts of alcohol (see Figure 6.5). The issues the students raised during the data creation discussion of this instructional activity indicate that they were beginning to *anticipate* limitations the data creation process that could influence the validity of conclusions that they could draw from the data. This episode also illustrates how the teacher and the students jointly built the data creation scenario through the questions that the students posed. This process of co-constructing data creation narratives was reasonably well established at this point in the eighth grade design experiment in that most students typically participated in asking the questions.

As a second illustration, consider an episode taken from one of the last instructional activities in the eighth-grade design experiment in which the students were to analyze two scatter plots that showed salary against years of education for women and for men, respectively (see Figure 6.9). The following excerpt is representative of the data creation discussion for this instructional activity:

1   Teacher: What I have here is data that shows the number of years of education that

2   different people have and how much money they make, what their annual salary is

3   So, if I looked at this person right here, this person had eight years of education, and

4   makes 40,000 dollars annually. So this person over here has 16 years of education

5   looks like they make about, maybe 12,000 dollars a year annually. Jerod?

6   Jerod: Does that count like elementary and high school and all that?

7   Teacher: Yes. Yeah.

8   Susan: Does that count college?

9   Teacher: Yeah. Total number of years of education. So this person only went

10   through—these people only went through the eighth grade. That's a very good

11   question Jerod. Brenda?

12   Brenda: What's like a normal amount of money to make a year?

13 Teacher: I don't know what a normal amount of money is …

14   Brenda: But like what would you consider good? Like, because me, I'm broke so the

15   5, how much is that, 50,000 dollars? That's pretty good to me.

As in the alcohol and reaction time episode discussed earlier, the students contributed to the joint development of the data creation narrative through the questions that they ask. Furthermore, these questions again indicate a need to understand the situation from which the data were generated. For example, in line 12, a student asked what a normal salary would be, indicating that she needed a point of reference from which to compare salaries of the men and women. During the subsequent data analysis discussion of this instructional activity, several of the raised concerns about the data creation process could jeopardize the conclusions that they had drawn from the data.

1   Susan: I have a question. Are these from like the same jobs, or from different jobs?

2   Teacher: You know, I had two really good…they're from different jobs. And that's a

3   really good question, and Jerod also asked a really good question when he was

4   this the first time whenever it was. Jerod asked me if these people had the

5   number of years experience at the job that they have. And we don't know that

6   information and that might be, both the thing you brought up Susan and the thing

7   Jerod has brought up are really important questions you would want to ask.

8   Susan: Because the person down there could be a janitor and the person up there

9   could be a doctor …

10 Teacher: But what they're saying, one of the questions that you're raising is there

11   might be more information you might want to find out about these people

12   What … that's right. Caitlin?

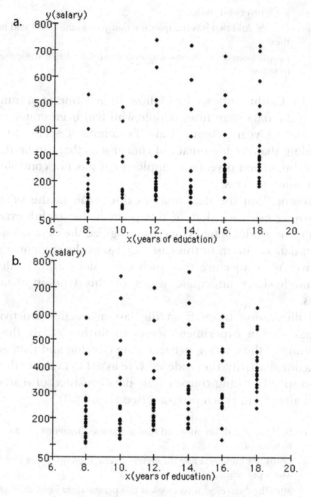

Figure 6.9.   a. Data for the years of education and salary activity showing women's years of education and their salaries. b. Data for the years of education and salary activity showing men's years of education and their salaries.

In raising the point that the data could have come from people of different jobs, Susan seemed to indicate that a comparison based on these data could be open to question because the men might have had jobs with inherently higher salary levels than the women.

13   Caitlin: People who went to like school who are, they could have went to school
                for

14   14 years of school but they're really high because they're like 50 and been
                doing their

15		job forever and stuff.
16	Teacher:	Exactly. And that was the question that Jerod said. Jerod said have all these
17		people been doing these—whatever these jobs are for the same number of years.

In line 13, Caitlin appeared to follow Susan's line of argument when she asked if the data were from people who had been employed for the same number of years. Again, Caitlin's comment seems to imply an understanding that the legitimacy of conclusions that can be drawn from the data is threatened if years-of-employment was not controlled during the data creation process.

This excerpt from the data analysis discussion of the salary activity provides strong evidence that, at this point in the eighth-grade design experiment, the students had come to recognize the dependency of conclusions on data creation. In this case, the issues the students raised were retrospective. We conjecture that such discussions are of value in that students might later anticipate issues of this type in data creation discussions.

A final illustration taken from the last instructional activity of the eighth-grade design experiment serves to further clarify the students' understanding of the relation between data creation and data analysis. In this instructional activity, the students were asked to evaluate the effectiveness of two speed-reading courses. The data consisted of reading speeds before and after taking each program (see Figure 6.3).

1	T:	So this is two different speed reading courses, remember? This is the number of
2		words per minute before the course … and then this is the number of words per minute
3		after the course. So if you look at this person right here, they were reading about 150
4		words before the course and now they're reading a little over 600 words. Yes?
5	Katie:	Do you know if they read the same thing?
6	T:	That's a very good questions. I don't know that.
7	Katie:	Cause that's not really fair. The person here, they're like 700 and when they're
8		done they read about 700
9	T:	Yeah, so it didn't do much good for them. So that's really a nice point
10		Katie's making. Katie is saying that this person right here didn't change much as a
11		result of the speed reading course.
12	Steve:	Do you think they might have taken it once before or something?

In line 5, a student asks a question pertaining to controlling variables—in this case, ensuring that all the participants in the study read the same text. Additionally, in line 12, a second student proposes a possible explanation for why someone's reading speed did not improve, namely because he or she had already taken the course before. These questions, which were representative of the questions asked during the remainder of this data analysis discussion, show a relatively high level of understanding of the data creation process. The students seemed to be aware of the consequences of failing to control for extraneous variables and were able to anticipate how aspects of the data creation process could influence the legitimacy of their conclusions.

The episodes we present here illustrate that the development of the data creation narrative had become a joint activity. Furthermore, the ways in which the students participated in this process indicate that they were concerned with the legitimacy of the conclusions that could be drawn from the data. This development would appear to have been made possible by the prior emergence of the students' role as data analysts. In particular, it is with respect to the purpose for generating data that certain aspects of a situation are seen to be significant and others as extraneous. Similarly, it is with respect to the purpose that a data set is judged to be representative or biased. As we demonstrate, it was as the students assumed the role of data analyst that they came to anticipate the importance of understanding the purpose for which the data were generated. The final shift we document in the students' understanding of the relation between data creation and data analysis involves interrogating the data creation process not only with respect to the ways in which they might structure the data but also with respect to the legitimacy of the conclusions they could draw from the data.

CONCLUSIONS

The research team did not consciously attempt to support the developments documented in this chapter when preparing for the two design experiments. The researchers were not, in fact, aware of these developments until midway through the eighth-grade experiment. As we note, the purpose of the data creation discussions was to ensure that the students were actually analyzing data rather than merely manipulating numerical values. Once this goal was achieved relatively early in the seventh-grade experiment, members of the research team focused on the primary objective of supporting the students' development of increasingly sophisticated ways of structuring data. As a consequence, the analysis we have presented is post hoc and describes the developments that were, from the research team's point of view, fortuitous.

We identify three ways in which the students' developing understandings of the relation between data creation and data analysis were supported. The first of these, the social norms established for data analysis discussions, gave the students license to ask clarifying questions during data creation discussions but did not orient them as to which aspects of the data creation process it might be important to interrogate. The second means of support, the sociomathematical norm of what counted as an acceptable databased argument, obliged the students to take account of the purpose for which the data were created when they conducted analyses. The third means of support involved the subsequent emergence of the students' role as data analysts. It was as the students assumed this role that they seemed to become aware of the importance of understanding both the purpose for which the data had been created and the ways in which their analyses would be used *before* they began an analysis. These developments made possible the final shift that we documented in which the students began to interrogate the data creation process not only with respect to the ways in which they might structure the data but also with respect to the legitimacy of the conclusions that they might be able to draw from the data. It was at this point that the students began to raise concerns about the control of extraneous variables and the representativeness of data sets.

The findings of this investigation have implications for both mathematics and science education. Recent reform documents such as the *Principles and Standards for School Mathematics* and the *National Science Education Standards* indicate that statistical data analysis is an important area of emphasis in both disciplines. The analysis we present indicates that it is not critical for middle school students to actually collect data for data sets to have a history that reflects the purposes for which they were generated. However, to reduce this study to a case of students coming to reason meaningfully about data that they have not generated themselves misses the larger point. As we illustrate, the teacher supported the students' developing understandings of the relationship between data creation and data analysis both by guiding the negotiation of social and sociomathematical norms, and by fostering the emergence of students' role as data analysts. This finding is potentially significant for mathematics teachers in that it suggests that they might be able to use limited instructional time more effectively by focusing on the process of data creation and the development of data-based arguments rather than on the actual collection of data. This possibility would also enable mathematics teachers to use data sets available on the Internet and other resources.

With regard to science education, this analysis indicates the importance of students talking through the data creation process even when they actually collect data. As we note, the research team initially thought of the data creation discussions as an alternative to data collection. It was only after the two experiments were completed that the team members came to view data collection as but one phase in the data creation process, one that involves actually making measurements. The science education literature indicates that students who are involved in collecting their own data often do not understand the fundamental reasons for doing so and are often more concerned with following laboratory protocols and getting "the right" data. As a consequence, "hands on" activities are often not "minds on" activities. Such cases might well be a consequence of skipping the phases of data creation prior to collection. Engaging students in discussions of the data creation process might enable them to remain cognizant of the purposes that underpin their inquiries and, eventually, to appreciate the influence of data creation on the legitimacy of the conclusions they can draw from the data they collect. In an approach of this type, data collection is seen to involve methodological and theoretical choices that are not transparent to students when they simply follow a laboratory protocol worksheet. It was these choices that became explicit topics of conversation in the two design experiments during the data creation discussions. The ways in which these discussions might need to be modified when students actually collect data is, of course, an issue for further investigation.

A second issue for future research—that is beyond the scope of our analysis but still pertinent to our findings—is that of students' interests. In the course of the two design experiments, the research team monitored the level of the students' engagement while developing and revising the instructional activities. These observations consistently indicated that the students were more engaged when instructional activities involved topics of broad social significance rather than of scientific significance. The teacher therefore appeared to be able to cultivate what might be termed the students' social interests but not their scientific interests. We tentatively speculate that students' engagement in investigations in which they actually collect data might be important to the cultivation of their scientific interests. In the absence of further evidence that bears on this issue, we recommend that mathematics teachers who adopt the approach of talking through the data creation process develop instructional activities that might potentially be perceived as socially significant by students.

ACKNOWLEDGMENTS

The analysis reported in this paper was supported by the National Science Foundation under grant No. REC9814898 and by the Office of Educational Research and Improvement under grant No. R305A60007.

The opinions expressed do not necessarily reflect the views of either the Foundation or OERI. In addition to the authors, the members of the research team for the design experiments were Jose Cortina, Lynn Hodge, Kay McClain, Kazu Nunokawa, and Nora Shuart. Cliff Konold and Erna Yackel served as long-term consultants and visited the classroom approximately once every two weeks throughout both experiments.

REFERENCES

Cobb, G. W., & Moore, D. S. (1997). Mathematics, statistics, and teaching. *American Mathematical Monthly, 104,* 801–823.

Cobb, P., & Whitenack, J. (1996). A method for conducting longitudinal analyses of classroom videorecordings and transcripts. *Educational Studies in Mathematics, 30,* 213–228.

Erickson, F. (1986). Qualitative methods in research on teaching. In M. C. Wittrock (Ed.), *The handbook of research on teaching* (3rd ed., pp. 119–161). New York: Macmillan.

Gravemeijer, K. (1999, April). *A rationale for an instructional sequence for analyzing one- and two-dimensional data sets.* Paper presented at the annual meeting of the American Educational Research Association, Montreal, Canada.

National Council of Teachers of Mathematics (NCTM). (2000). *Principles and standards for school mathematics.* Reston, VA: National Council of Teachers of Mathematics.

National Research Council (NRC). (1996). *National science education standards.* Washington, DC: National Academy Press.

Roth, W. -M. (1996). Where is the context in contextual word problems?: Mathematical Practices and products in Grade 8 students' answers to story problems. *Cognition and Instruction, 14,* 487–527.

CHAPTER 7

HOW DO YOU KNOW WHICH WAY THE ARROWS GO?

The Emergence and Brokering of a Classroom Mathematics Practice

Chris Rasmussen, Michelle Zandieh, and Megan Wawro

In this chapter we analyze how a particularly rich and complex inscription known as a bifurcation diagram emerged in an undergraduate differential equations class. We frame the analysis in terms of the construct of a classroom mathematics practice and demonstrate how the brokering moves of the teacher and some students functioned as a mechanism for the initiation and ongoing growth of this practice. Our analysis contributes more broadly to contemporary accounts of how creating, using, and interpreting inscriptions is a social process that reflects the evolving culture of the classroom community, as well as the culture of the broader mathematical community.

The chapter is divided into three main sections. The background section provides a brief survey of the literature on representations, followed by a discussion of what is meant by a classroom mathematics practice and by brokering. The results and discussion section consists of four parts that

Mathematical Representation at the Interface of Body and Culture, pp. 171–218
Copyright © 2009 by Information Age Publishing

chronologically analyze students' reasoning over the course of two class sessions. In the conclusion section, we detail the three broad categories of broker moves that are identified through our reflection on the various broker moves that serve as a mechanism for the emergence of a bifurcation diagram.

COMMUNITIES OF MATHEMATICAL PRACTICE

The emerging classroom mathematics practice that we analyze in this chapter centers on students' (re)invention of a fairly complex representation known by experts as a bifurcation diagram. To situate our analysis within the vast literature on representations, we begin with a brief review of some early work that was dominated by cognitive framings and then point to more recent work informed by contemporary social and cultural perspectives.

A Social Practice Perspective on Mathematics

In a move away from the traditional representational view of mind approach, Roth and McGinn (1998) replace the problematic notion of representation with the term inscription, which refers to "signs that are materially embodied in some medium, such as paper or computer monitors" (p. 37). Examples of inscriptions are tables, lists, diagrams, graphs, equations, and spreadsheets. Because of their material embodiment, inscriptions are accessible by all learners, as opposed to mental representations, which are inaccessible to all members of a learning community except the particular individual whose mental representations might be under consideration. Roth and McGinn, therefore, assert that by focusing on the development, transformation, and use of graphical inscriptions in a social setting rather than on the activity of an individual mind, we are presented with a richer framing through which to understand learners and graphical symbolism. Our analysis in this chapter aims to contribute in this direction.

This shift in focus coincides with an increased interest in theoretical and pragmatic accounts of learning and teaching from cultural and sociological points of view (Wenger, 1998). A common thread among this more recent work is a focus on how communities of learners participate in and develop various classroom mathematics practices that tend to become normative with respect to how students in a classroom community reason, symbolize, and argue. Such practices are established through patterns of interaction center on specific forms of mathematical activity. For example, one of six practices emerging in an undergraduate mathematics course in differential equations was referred as the "creating and structuring a slope field practice" (Stephan & Rasmussen, 2002). This particular classroom mathematics practice entailed three specific ways of reasoning: detailing the way in which slopes change over time, justifying why slopes are invariant horizontally for autonomous differential equations, and imagining an infinite number of tangent vectors when only finitely many are visible. As this example illustrates, these researchers characterize a classroom mathematics practice in terms of a *cluster* of related forms of activity. This characterization grows out of yet differs from earlier characterizations put forth by Cobb and Yackel (1996) in that the early characterization of a classroom mathematics practice involved one form of mathematical activity, rather than a cluster. Classroom mathematics practices typically evolve over several days and even weeks. As such, classroom mathematics practices are not pre-established forms of activity into which students are inducted, but rather emerge as participants interact. Hence, different classrooms using the same curriculum can develop a somewhat different collection of classroom mathematics practices with somewhat different forms of activity.

The nature of classroom mathematics practices involves two additional characteristics. First, classroom mathematics practices can be established in a non-sequential temporal order. In a previous analysis that documented the constitution of classroom mathematics practices in a first-grade mathematics class (Stephan, Bowers, Cobb, & Gravemeijer, 2003), the various practices proceeded in a more or less sequential fashion: the initiation and constitution of the first practice preceded the initiation and constitution of the second practice in time. Such temporal orderliness is not always the case (Stephan & Rasmussen, 2002). Second, classroom mathematics practices can emerge in a non-sequential structure. That is, one or more of the specific forms of activity that make up one practice can also be part of a different practice. For example, in creating and structuring a slope field practice, the form of activity referred to as justifying why slopes are invariant horizontally for autonomous differential equations might very well appear in an entirely different classroom mathematics practice. The analysis presented in this chapter takes this structural overlap even further by suggesting that an entire classroom mathematics practice can become embedded in another practice.

Evolving classroom cultures tend to be mediated by the broader culture of the discipline of mathematics. This is because the goals and values of the classroom participants, as well as the curricular materials, are inseparable from the discipline itself. Teachers and texts do not exist in isolation from the practices of the broader field of mathematics. The practices of this broader culture can be characterized in terms of the more generic activities of defining, algorithmatizing, proving, modeling, and symbolizing (Rasmussen, Zandieh, King, & Teppo, 2005). The classroom mathematics practice that we analyze in this chapter is a particular manifestation of the discipline practice of symbolizing. Mathematicians routinely create, interpret and use inscriptions to solve problems, communicate with colleagues, convince others, and so forth. Thus, particular classrooms, especially those that are inquiry-oriented with the intent of engaging students in the authentic practice of mathematics, might engage learners in one or more of these broader discipline practices. In this way, students' participation in a local classroom mathematics practice is a way in which they are enculturated into the cultural practices of mathematics.

Brokering and the Emergence of Mathematics Practices

In this chapter we offer insights into the mechanisms by which classroom mathematics practices are initiated and begin to take hold. To accomplish this we analyze in depth two successive class sessions. We demonstrate how brokering functions as a mechanism that supports the initiation and emergence of one particular classroom mathematics practice in which students engage in symbolizing. Our use of the term brokering draws on the work of Wenger (1998). Brokering is a mechanism that influences the degree of continuity between communities. In our analysis, we consider three different communities: the broader mathematical community, the local classroom community, and the various small groups that make up the local classroom community. As we detail in this chapter, the brokers in these communities are the teacher and specific students in the class. A broker is someone who has membership status in more than one community. For example, in our case the teacher is a legitimate peripheral member of the broader mathematics community, a full member of the classroom community, and a peripheral member of each of the small groups that make up the classroom community.

The job of brokering is complex (Wenger, 1998). Brokers facilitate the translation, coordination, and alignment of perspectives between communities. Brokers must be able to mobilize attention and address different points of view. Brokers are unique in that they are "able to make new connections across communities of practice, enable coordination, and—if they are good brokers—open new possibilities for meaning" (p. 109). Wenger also argues that the job of brokering requires the ability to "cause learning" by introducing into one community elements of practice from a different community (p. 109). For example, the teacher as broker might introduce formal terminology from the discipline of mathematics into the classroom community. On the other hand, particular students who are adept at certain forms of activity that make up a classroom mathematics practice might act as brokers to facilitate others in the class as they move from periphery to more central participants in this practice. This latter example is brokering between different communities within the same classroom. In other words, learning is evidenced by the initiation and evolution of classroom mathematics practices, and it is toward this end that we now turn.

A CASE STUDY OF BROKERING

In this section we detail the initiation of a classroom mathematics practice (CMP) that involved the creation and interpretation of a bifurcation diagram and the brokering moves that functioned as a mechanism for the emergence and appropriation of this practice. We refer to this particular CMP as the "summarizing the changing structure of the solution space practice." By structure of the solution space, we mean the characteristics of the collection of functions that satisfy the differential equation. These characteristics include the number, location, and type (sink, source, or node) of equilibrium solutions, the concavity of solution graphs, and the horizontal invariance of graphs of solution functions to autonomous differential equations.[1] In our analysis, we demonstrate how the teacher, together with the students from one particular small group, functioned as brokers for others in the class to interpret, understand, and participate in this particular practice.

Data for this analysis comes from a 15-week classroom teaching experiment conducted in an undergraduate differential equations course. Data sources consisted of daily classroom video recordings from two cameras, video recordings of student interviews, and copies of student work. The teaching experiment was conducted as part of a larger research program aimed at developing an inquiry-oriented, research-based instructional approach in undergraduate mathematics.

Our analysis focuses on data collected from two consecutive classroom sessions. We find that this particular CMP entails five related ways in which students worked with various inscriptions and related symbolic equations. In particular, the summarizing the changing structure of the solution space practice consisted of five forms of mathematical activity (Figure 7.1). The forms of activity figure prominently in the analyses presented in the subsequent discussion. We determined these five forms of activity retrospectively through repeated examination of the data, but present them here prospectively to facilitate organization and discussion of results.

(a) Reasoning with single dy/dt versus y graph to ascertain the structure of the solution space.

(b) Reasoning with discretely or continuoulsy changing dy/dt versus y graphs to ascertain the changing structure space.

(c) Solving the equation $dy/dt = 0$ and interpreting the result in terms of the equilibrium values as a function of some parameter.

(d) Plotting the equilibrium solutions (values) as they vary as a function of the paramenter.

(e) Re-presenting the structure of the solution space with phase lines embedded in the equilibrium value versus parameter graph.

Figure 7.1. Summarizing the changing structure of the solution space practice

We present our study in four parts. In each of the four parts we emphasize the brokering moves that function as a mechanism for the emergence of the summarizing the changing structure of the solution space practice. In addition, in the first part we describe the initial task that eventually led to some students creating a bifurcation diagram and detail students' initial work on it. In the second part we detail the presentation of Brady's small group and point to ways in which their presentation was similar to or different from the forms of mathematical activity that constitute the summarizing the changing structure of the solution space practice. In the third part we describe how Kenneth and Lorenzo's small group presentation essentially reinvented a bifurcation diagram and how their presentation relates to the various forms of activity that constitute the changing structure of the solution space practice. In the final part we return to Brady's group to detail their work on a new task and their transition to becoming more central participants in the summarizing the changing structure of the solution space practice.

Part 1: Introduction of and Initial Work on the Fish.com Task

In this section we detail the initiation and start of a problem that eventually led to one group of students creating what an expert would recognize as a bifurcation diagram. To emphasize, the bifurcation diagram these students invented was the product of their own creative activity and not something that was appropriated from a textbook, a teacher, or some other resource. Moreover, the creation, interpretation, and further rendering of this inscription entailed a number of different forms of mathematical activity (Figure 7.1). It is these various forms of mathematical activity that constitute the summarizing of the changing structure of the solution space practice.

In the first of the two sessions analyzed here, the teacher introduced the "fish.com" problem (Figure 7.2). After a brief discussion of the problem statement, the teacher discussed with students the need to modify the differential equation $\frac{dP}{dt} = 2P\left(1 - \frac{P}{25}\right)$ in some way because the given differential equation does not take into account harvesting by the local public. Before trying to decide on an appropriate modification, however, the teacher requested that students work in their small groups to first fully understand what the rate of change equation $\frac{dP}{dt} = 2P\left(1 - \frac{P}{25}\right)$ predicts for the fish population for various initial conditions. After completing their analysis, the teacher invited two students from different small groups to present their findings. Figure 7.3 shows the different inscriptions that these students used in their analysis.

In Figure 7.3a, Roy presents a sketch of a slope field, which he said was generated using his graphing calculator, and the corresponding flow line (phase line) to the right of the slope field. The features of the graphs that Roy chose to discuss included how to draw the flow line from the slope field, the existence of equilibrium solutions at $P = 0$ and $P = 25$, and the conclusion that solutions with positive initial conditions "will all just kind of go towards 25." Roy's carefully drawn slope field strongly suggests horizontal invariance in the tangent vectors and the concavity for various solutions. Although negative P values do not make sense for this situation, students in this class had been accustomed to analyzing the full range of solutions for the sake of mathematical completeness. Roy's graphs reflected such analyses.

A scientist at fish hatchery has previously demonstrated that the rate of change equation $\frac{dP}{dt} = 2P\left(1 - \frac{P}{25}\right)$ is a reasonable model for predicting the number of fish that the hatchery can expect to find in their pond.

Recently, the hatchery was bought out by fish.com and the new owners are planning to allow the public to catch fish at the hatchery (for a fee of course). The new owners need to decide how many fish per year they should allow to be harvested. Prepare a report for the new owners that illustrates the implications that various choices for harvesting will have on future fish populations.

Figure 7.2. Fish.com problem.

Figure 7.3.   Student analyses of $\frac{dP}{dt} = 2P\left(1 - \frac{P}{25}\right)$.

In Figure 7.3b, Kurt presented similar conclusions about the solution functions $P(t)$, but his work utilizes a graph of dP/dt versus P rather than a slope field. Based on what happened during the two lessons, it appears that interpreting such graphs to ascertain the structure of the solution space was in and of itself a distinct CMP. This finding is consistent with other findings in a different, but similarly taught differential equations class (Stephan & Rasmussen, 2002). In reference to Figure 7.1, this distinct practice becomes embedded in the summarizing of the changing structure of the solution space practice. Because use of such graphs will figure prominently in the subsequent sections, we provide transcript and commentary on how Kurt (and his group) interpreted and reasoned with such inscriptions.

In his presentation, Kurt pointed back and forth between the graph of dP/dt and the flow line immediately to the right of the parabola shown in Figure 7.3b.

> Kevin: Greater than 25 [points to the graph of dP/dt versus P where $P > 25$] you're going to have a negative [points to the portion of the flow line above $P = 25$ with a downward pointing arrow]. Between 0 and 25 [points back to the graph of dP/dt versus P where $0 < P < 25$] you're in the positive [points back to the portion of the flow line with an upward pointing arrow], when it's less than zero [points back to the graph of dP/dt versus P where $P < 0$] you're going to have a negative [points back to the flow line where $P < 0$].

Kurt's interpretation of the dP/dt versus P graph focused on the regions in which the sign of dP/dt is positive or negative. For example, for P between 0 and 25, Kurt stated that "you're in the positive," meaning that the graph is *in* the region above the P-axis. Similarly, when the graph of dP/dt is below the P-axis the rate of change is negative and hence the flow line would point downward to indicate decreasing solution functions. This is consistent with how students in the class interpreted other graphs of autonomous differential equations. Next, Kurt explained how he then drew representative graphs of the functions that solve the differential equation.

> Kurt: And from there you can just draw like basic equations, generalizations of it [points to the flow line between 0 and 25]. So if you start anywhere between 0 and 25, it's [points to the graph of P versus t and begins to trace out the solution graph as he talks] going to increase slowly by zero and then more rapidly in the middle and then slower again at 25. It's going to decrease to 25 [tracing out the graph of P versus t above 25]. Same thing for the opposite [i.e., when P is less than 0] but you can't have a negative fish population, so it's all going to go towards 25.

Although Kurt did not elaborate on what he meant by "basic equations," students' prior work in this class suggests that he might have meant that the graphs he drew were "basic" in the sense that other solution graphs could be obtained from these simply by shifting the graphs drawn left or right. This result follows from the fact that the differential equation is autonomous, and is consistent with Roy's slope field that shows tangent vectors with a horizontal invariance to their slopes. Kurt also noted that, except for negative fish populations, "it's all going towards 25." In other words, the equilibrium solution of 25 attracts nearby solutions. Students in this class referred to such equilibrium solutions as a "sink."

Kurt's explanation involved a significant amount of pointing back and forth between the various inscriptions, which implicitly brought forth connections between the inscriptions. Neither the teacher nor other students requested additional explanation or clarification for how he came to his conclusions and for why these conclusions were valid. The teacher even provided ample opportunity for students to ask Kurt questions. For example, after Kurt finished explaining, the teacher commented as follows:

> Teacher: So both Roy and Kurt it seems to me have the same conclusions but slightly different ways of going about organizing or getting that information. Good. Ok, let's get, are there any questions on this? Folks want to ask them questions, clarifying questions, or bring up some more points they want to be more picky on? [14 second pause] Everyone ok then? Ok, ok let's go to the—thanks, guys—so, so the next part of the problem.

In inquiry-oriented classrooms such as this one, the lack of or dropping off of challenges or requests for justification is a sign that particular ideas and interpretations are functioning "as if" the classroom community shared them (Rasmussen & Stephan, 2008). In this particular class, students routinely indicated disagreement or asked questions when in doubt. However, they did not do so in this case, which suggests that the ideas related to Kurt's presentation functioned as if they were shared by others in the class. In other words, it appears that reasoning with a graph of dy/dt versus y had become relatively routine for these students. Indeed, Kurt's use of a dP/dt versus P graph was not the first time that students had used such an inscription to glean information about the structure of the solution space. For example, in a previous class session students had been given a graph of an autonomous differential equation dN/dt as a function N and had used this graph to ascertain the structure of the solution space.

In terms of the reasoning about the changing structure of the solution space practice, Kurt's interpretation and use of the graph of dP/dt versus P is listed as aspect "a" in Figure 7.1. As stated earlier, we see this particular form of activity as a distinct CMP. While a full accounting of the forms of activity that constitute "reasoning with a single dy/dt versus y graph to ascertain the structure of the solution space" is beyond the scope of this chapter, we surmise from Kurt's presentation that this CMP includes the following forms of activity: (a) interpreting the y axis intercepts of a dy/dt versus y graph as equilibrium solutions; (b) determining whether solution functions are increasing or decreasing by graphically examining the sign of the derivative; (c) labeling the equilibrium solutions (e.g., sink or source) by the behavior of nearby solutions; and (d) making connections between a graph of dy/dt versus y, a phase line, and solution graphs in the y versus t plane. All of these activities were evident in Kurt's presentation, and are part of Parts 3 and 4 of this section.

After Roy and Kurt's presentation, the teacher then requested that students try to modify the rate of change of equation $\dfrac{dP}{dt} = 2P\left(1 - \dfrac{P}{25}\right)$ so that it would account for the new owners' decision to allow the public to catch fish. For approximately 6 minutes students brainstormed various ways to modify the differential equation in their small groups. This was followed by a 14-minute whole class discussion of students' ideas and rationales. This discussion served to bring out ideas from the small groups and to cultivate new ideas that emerged for some students as their classmates explained their thinking. The different ideas that were discussed in whole class were to subtract from the given differential equation the following: a constant k, some function of time $h(t)$, the term $k\left(1 - \dfrac{P}{12.5}\right)$, and a fraction of the current population, for example $\dfrac{1}{3}p$. The rationales for these ideas that students gave were quite sophisticated, and the teacher noted the many sensible ideas for modifying the differential equation. In the end, however, the teacher decided that the class was to use the following modified differential equation: $\frac{dP}{dt} = 2P\left(1 - \frac{P}{25}\right) - k$, where k represents a constant annual harvesting rate. The teacher justified this choice to students as follows: "So the reason I'm doing this [choosing the $-k$ modification] is because I didn't want 10 different groups having different differential equations because then we don't have a common basis to talk about our analyses." Although the teacher did not mention this to the class, we note that this particular modification, in comparison to the other suggested modification, lends itself more readily to discretely or continuously changing the original dy/dt versus y graph.

Teacher Broker Moves

In the previous paragraphs we outline the various forms of activity that constitute the summarizing the changing structure of the solution space practice and begin detailing students' mathematical reasoning related to these forms of activity. Next, we focus our attention on the teacher to highlight the unique broker moves as they played out in Part 1. Even in the relatively short amount of class time that transpired in Part 1, the teacher played a number of distinct roles as a broker between the classroom community and the mathematical community, and between the classroom community as a whole and individual small groups.

First, the teacher played a critical role as broker between the classroom community and the broader mathematical community in terms of task selection and the way in which he engaged students in the task. Specifically, it was the teacher who selected the fish.com task because it had the potential for student creation and reinvention. In other words, it is the teacher who is in a position to recognize characteristics of tasks that are likely to be productive for enabling newcomers to be more central participants in the discipline of mathematics, such as modeling and symbolizing. For example, the fish.com problem was recognized by the teacher as a task he could use to engage students in modeling by revising a differential equation to fit new assumptions and analyzing this new model. The fish.com problem, as used by the teacher, also engaged students in symbolizing by virtue of the fact that the task invited them to develop "a report for the new owners that illustrate the implications that various choices for harvesting will have on future fish populations." The teacher did not prescribe what form this report was to take and hence allowed students to develop their own inscriptions for the owners. At the same time the teacher opened up the task in terms of not prescribing methods for analysis or ways to present findings, the teacher also set constraints and boundaries on the task. In particular, the teacher emphasized to the class that their job was not to tell the owners what harvesting rate to use, but rather to provide information to the owners so that they could make an informed decision. The teacher also constrained presentations to a single overhead transparency to encourage students to figure out a way to consolidate their analysis. As such, the teacher's task selection and the way in which he engaged students in the task provided the opportunity for him to act as a broker between the discipline of mathematics and the way in which the classroom community engaged in these activities.

Second, the teacher played a critical role as broker between the classroom community and different small groups within this community. One particularly noteworthy brokering move occurred when the teacher invited two different students, Roy and Kurt, to share their analysis of the original differential equation with the whole class and others to comment or ask questions about the presentations. The significance of this broker move is that it resulted in situating the new problem within previously established practices in a way that did not privilege particular analysis techniques nor prescribe how to proceed. For example, after Kurt's presentation the teacher commented, "So both Roy and Kurt it seems to me have the same conclusions but slightly different ways of going about organizing or getting that information. Good." It also encouraged ownership of analyses, positioned students as capable of making progress, and facilitated connections across approaches.

All of the brokering moves that occurred in Part 1 served the purpose of creating the background that allowed for the emergence of the summarizing the changing structure of the solution space practice. Moreover, these brokering moves helped create a situation for subsequent brokering moves by the teacher for others in the class to more fully participate in this practice. In Parts 2–4 below we fully develop these points.

Part 2: Brady's Presentation of the Fish.com Task

In this part, we discuss and analyze Brady's presentation of his group's work on the modified differential equation $\frac{dP}{dt} = 2P\left(1 - \frac{P}{25}\right) - k$ in the fish.com task. Brady's presentation relied heavily on reasoning with a table of harvesting rates and population values, and he made no explicit references to graphs of dP/dt versus P. He was able to engage in conversation with the class about his group's ideas, explain examples as well as connect to mathematical concepts not initially mentioned in his presentation. In Episode 2a, we detail Brady's presentation of his group's work. In Episode 2b, we discuss various ways in which the teacher and another student served as brokers between Brady and the rest of the class as they facilitated the clarification of Brady's approach as well as the connection to previously established mathematical ideas. We conclude this section with a discussion in which we point to ways in which Brady's presentation was similar to or different from the forms of mathematical activity that constitute the summarizing of the changing structure of the solution space practice. In particular, we argue that Brady's presentation offered a tabular way to summarize the changing structure of the solution space that utilized reasoning with the dy/dt equation in a way that was more algebraic and numeric, in comparison to Kurt's presentation in Part 1 and to what we will see in Part 3. In addition, we claim that Brady's presentation provides evidence that he and his group were poised to participate in the emerging practice, which we will detail in Part 4.

Episode 2a: Brady's Presentation

Brady was the second person to present a small group's work on the modified differential equation $\frac{dP}{dt} = 2P\left(1 - \frac{P}{25}\right) - k$ in the fish.com task.

As we see from Brady's presentation, the table served as the group's main tool for summarizing the situation and making decisions about harvesting rates in the required 1-page form (Figure 7.4). The table's left-most column contained the values that Brady and his group used for the harvesting rate H (his group used "H" instead of "k" as the variable for the harvesting rate parameter). Notice that these H-values are integers, ranging from one to twelve. The other two columns, labeled "POP MAX" and "POP MIN," contained the values that bound the range of the maximum and minimum populations that are associated with each particular harvesting rate that yield a positive rate of change.

Brady began his presentation by stating his group's initial thinking that it would be useful or "ideal" to maintain a positive growth rate.

> Brady: Okay, um, when we did ours, um, our idea was that since you have a hold-
> ing factor of basically twenty-five, if you go over twenty-five you automat-
> ically start going down. So we said that it was ideal to keep your
> population between zero and twenty-five. So we kind of assumed that.

Brady then detailed how they used the equation, case by case for integer H-values between one and twelve, to fill in the POP MAX and POP MIN columns in their table: "Then we used our [differential] equation and we found that we had a max population and a minimum population for each H-value. So for every H-value if you had a population between this range, that you could go, pulling this many fish." Brady did not explain how he and his group used the differential equation to find the

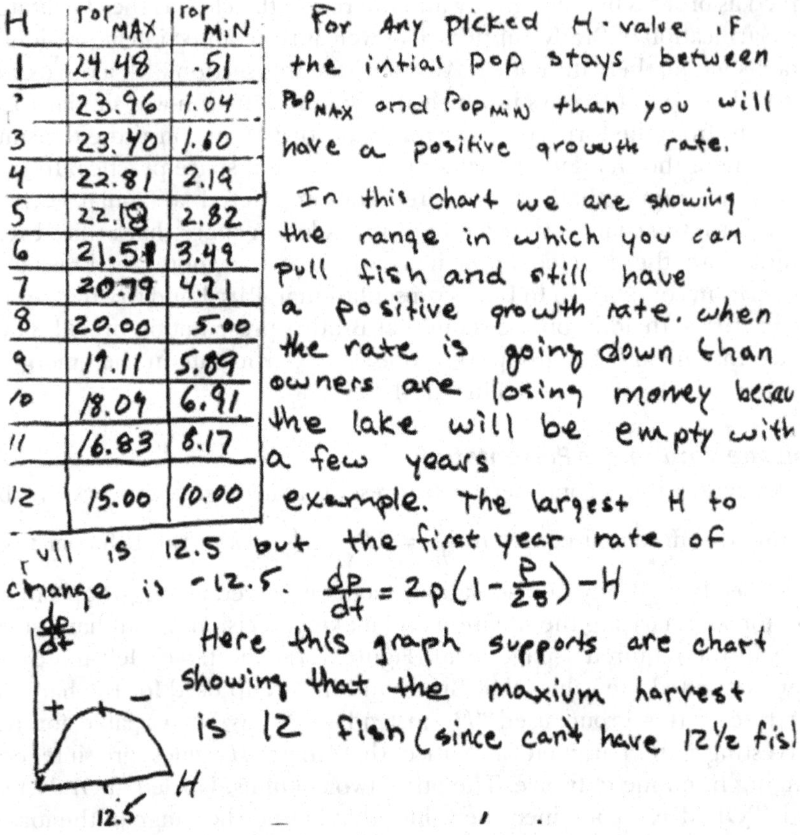

The handwritten overhead:

H	POP_{MAX}	POP_{MIN}
1	24.48	.51
2	23.96	1.04
3	23.40	1.60
4	22.81	2.19
5	22.18	2.82
6	21.51	3.49
7	20.79	4.21
8	20.00	5.00
9	19.11	5.89
10	18.09	6.91
11	16.83	8.17
12	15.00	10.00

For any picked H·value, if the intial pop stays between POP_{MAX} and POP_{MIN} than you will have a positive growth rate.

In this chart we are showing the range in which you can pull fish and still have a positive growth rate. when the rate is going down than owners are losing money becau the lake will be empty with a few years

example. The largest H to pull is 12.5 but the first year rate of change is −12.5. $\frac{dP}{dt} = 2p(1 - \frac{P}{25}) - H$

Here this graph supports are chart showing that the maxium harvest is 12 fish (since cant have 12½ fis)

Figure 7.4.   The overhead for Brady's group.

corresponding population values for each H-value. However, because the table has each population value given to two decimal places (Figure 7.4), we strongly suspect that the group solved for the values of P that were solutions to $2P\left(1 - \frac{P}{25}\right) - H = 0$ for each of the table's given H values.

Next, Brady illustrated how to interpret their table by taking the class through an example from the table. In particular, Brady asked the class to imagine that their pond had an initial population of twenty-one units of fish. He then found where twenty-one would be situated in the range of values given in the POP MAX and POP MIN columns and concluded that the chart's corresponding H-value of six (Figure 7.5) would be the best amount of fish to pull.

Brady: Like, uh, if you were gonna, if you had a population of 21, then, uh, you could say, okay, well 21 is between, uh, 21.5 one and 3.49, so your best H to hold would be 6. And the whole point of these numbers are, is, between these two numbers, when you pull that many fish, then you're going to have a positive rate of change. So you'll have an increase in the fish population, and we thought that, uh, an increase in the fish population here was an important thing for a fish company, because then they wouldn't have to restock their fish.

Brady's rationale for the "best H" corresponding to an initial population of 21 hinged on the fact that the rate of change between POP MIN and POP MAX is positive and hence, as Brady made explicit, "you'll have an increase in the fish population." Neither the teacher nor other students asked for clarification on how Brady and his group determined that the rate of change was positive. Perhaps he and his group determined the sign of dP/dt by inserting specific P-values into the differential equation, or perhaps they created slope fields similar to what Roy did (Figure 7.3a). Although we cannot be certain how they determined the sign of dP/dt, we think that it is highly unlikely that Brady and his group used a graph of dP/dt versus P for this purpose. Evidence for this interpretation comes from the fact that although a dP/dt versus P graph appears in the bottom left corner of their overhead (Figure 7.4), this graph was never mentioned or discussed by Brady. Moreover, the axes on the graph are incorrectly labeled. Thus, it is unlikely that Brady's group used this graph as a reasoning tool. Further evidence for this claim are detailed in Part 4.

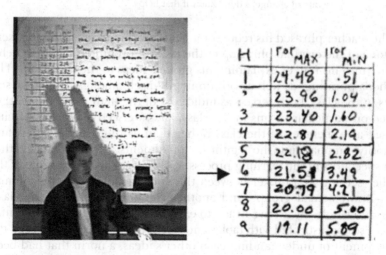

Figure 7.5.   Brady explains how to use their chart.

Brady's explanation of how to interpret their table indicates his attention to the structure of the solution space. Recall that by "structure of the solution space," we mean the characteristics of the collection of functions that satisfy a differential equation, which includes the number, location, and type (sink, source, or node) of equilibrium solutions. For example, Brady used an initial population of 21 as a generic example to explain how to read their table. Using language closely tied to the context, his explanation essentially detailed how POP MAX functions as a sink and the POP MIN functions as source, thus indicating the structure of the solution space. Brady's ability to reason about the structure of the solution space will be revisited in the discussion, where we discuss how Brady's presentation relates to the summarizing the changing structure of the solution space practice.

Episode 2b: Teacher and Student Broker Moves

A few moments after Brady's explanation, the teacher—who was sitting down in a desk among the other students in the class—prompted Brady to explain his group's table a bit more.

> Teacher: Brady, let's pretend that I'm an owner and I make a lot of money, and I'm an executive, but I'm not so sophisticated with tables and stuff. So I need some help understanding that table that you've got up there.
>
> Brady: Oh, okay. Alright, for each H-value, that's how many fish you can pull [he points at the table] and for each one, if you go across, the population max is—and the population min—that is the range of populations; that's, uh, so it's the highest amount of fish and the lowest amount of fish that you can have in your lake, pull this many fish out, and still get a positive rate of change. I don't know if that helps.

The teacher phrased his request for clarification in such a way that kept the tone of the original phrasing of the task yet prompted Brady to delve further into his role of explaining his group's ideas to the classroom. The teacher's phrasing, as well as positioning himself amongst the other students in the class, both served as indicators of the high value the teacher placed on particular norms in the class. First, we see the teacher placing himself as a member of the class rather than as the outside expert and authority, emphasizing the legitimacy of students as participants rather than spectators in the learning process. We also see this an example of the teacher serving as a broker between the understandings of one community—Brady's small group—and another community—that of the classroom as a whole. By asking Brady to explain again for someone who may not be "so sophisticated with tables and stuff," the teacher encouraged the development of understanding each other's ideas, a norm that had been in development since the semester began. Brady concluded with, "I don't know if that helps," which implies that he realized his explanation was supposed to serve the purpose of further clarifying his group's approach. This claim is supported by the video, which shows Brady looking around the room, rather than only at the teacher, who initially requested more explanation.

After Brady's explanation, which ended with "I don't know if that helps," the teacher followed up by acting in accordance with the norm that students listen to and respond to their classmates' questions or requests.

Teacher:	[7 second pause] So that's a question to you all—does that explain what his table is? If you're the owners, are you understanding what he's, what information he's provided? Nathan—kind of?
Nathan:	Well, if I didn't know anything about math, you know, like, then probably not. But I understand what he's saying.
Teacher:	Summarize for us, Nathan, your interpretation of the information he's presenting.
Nathan:	He's saying that the max and min values are like the equilibrium points, where if you're in that range and you pull that amount of fish, it will be increasing towards the max equilibrium point.
Brady:	Yeah, and if you are exactly on the number then you would have a zero rate of change.
Nathan:	Right, it'll just stay there.

The teacher waited for seven seconds before he spoke, which allowed sufficient time for students to contribute their response if they had one. When no class member spoke, the teacher did not comment on the potential effectiveness of Brady's explanation; rather, he directly requested that the students respond as if they were the owners of fish.com. This short exchange shows the benefit of the teacher acting in accordance with the norm that students listen and respond to their classmates' ideas. In particular, as a result of the teacher's actions, Nathan in essence acted as a broker to connect Brady's group terminology of POP MIN and POP MAX to the conventional terms of equilibrium points. When asked to summarize his interpretation of Brady's work, Nathan stated, "The max and min values are like the equilibrium points." Brady immediately agreed and added more interpretation about what it meant to have a zero rate of change. Brady's agreement indicates that Nathan had interpreted the results as he had intended.

After a bit of discussion about the contextual sensibility of particular population values and their corresponding H-values, the teacher directed Brady's attention back to Nathan's assertion that POP MIN and POP MAX were equilibrium solutions.

Teacher: Oh, does it, is that [POP MIN of 1.04] an equilibrium? I think I heard Nathan say—

Brady: I think it is an equilibrium point.

Teacher: Are those population min and population max, are those supposed to be equilibrium solutions?

Brady: Yeah, yeah. If your population is 1.4 [sic 1.04] and you pull 2 fish a year, then your population will stay at 1.04.

We see the teacher's statement "Is that an equilibrium? I think I heard Nathan say—" as serving two purposes. First, explicitly attributing ideas to students enforces a sense of ownership of ideas among the students, which in turn helps students develop a view of themselves as capable thinkers and doers of mathematics. Second, the teacher called attention to Nathan's role of brokering in this situation after the latter had previously "translated" Brady's explanation into the notion of equilibrium points. This was a well-established idea by this stage of the semester for this class. Finally, the teacher himself served as a broker here as well, choosing to further emphasize the association between what Brady was describing contextually and a previous mathematical idea.

Connections to the Mathematical Activity of the Emerging Practice

Brady's presentation suggested a tabular way to summarize the changing structure of the solution space that utilized reasoning with the dP/dt equation in a way that was more algebraic and numeric, in comparison to the graph-oriented reasoning we see in Kurt's presentation in Part 1 and in what we see in Part 3. That is, Brady's use of the dP/dt equation constitutes a separate but complementary mathematical activity than those listed as aspects "a" and "b" (Figure 7.1). In Part 1 we see Kurt reason with the dP/dt versus P graph to ascertain the structure of the solution space (aspect a) of the emerging practice, and in Part 3 below we see how another group presentation utilized dP/dt versus P graphs with varying H-values in order to ascertain the changing structure of the solution space, aspect "b" of the emerging practice. Brady and his group's table-based reasoning, while different than either of the aforementioned graph-related activities, provided both affordances and limitations for becoming more central participants in the emerging summarizing the changing structure of the solution space practice. Here we detail what affordances this way of thinking provided Brady and his group, and in Part 4 we return to what limitations they encountered by not reasoning with the dP/dt versus P graphs.

Brady's group did not appear to utilize to any large degree a graph of dP/dt versus P. Hence we see little evidence that they engaged in the forms of activity a) or b) from Figure 7.1, both of which capitalize on graphs of

dy/dt versus y. However, the group's way of reasoning, even without the graphs, aligned them to some degree with the underlying reasoning behind certain aspects of the emerging practice. Here we suggest that the group's table-based reasoning aligns them, at least partially, with the reasoning that is manifested in graphical form in activities "c," "d," and "e" of the emerging practice (Figure 7.1).

When explaining his group's approach, Brady stated in Episode 2a, "Then we used our equation and we found that we had a max population and a minimum population for each H-value." Although Brady did not explicitly use the terms "equilibrium point" or "equilibrium solution" in his presentation, he quickly agreed with both Nathan and the teacher when they used one of these terms in place of "population max" or "population min." The above quote, along with Brady's quick agreement with Nathan and the teacher about the terminology, are evidence that the tabular-based reasoning exhibited by Brady's group was well-aligned with the graph-oriented reasoning elicited in aspect "c" of the emerging practice. In addition, this is also true in regard to aspect "d." Brady and his group "plotted" the information in a table rather than in a graphical format. Although Brady's table presents only partial information about the relationship between the parameter and equilibrium solutions, the underlying reasoning behind this tabular presentation was very well-aligned with the reasoning that is manifested graphically in aspect "d" of the emerging practice.

Finally, Brady's presentation shows evidence that his group was in position to participate in the form of activity "e." Although Brady's group did not draw or discuss phase lines, his explanation of how to interpret their table included much of the same information that a graph of the equilibrium values versus parameter with phase lines embedded would have offered. Brady had said, "Alright, for each H-value ... it's the highest amount of fish and the lowest amount of fish that you can have in your lake, pull this many fish out, and still get a positive rate of change." The group therefore was already reasoning, at least in part, about the changing structure of the solution space. This information, provided verbally through Brady's explanation, provides essentially the same information that would be given by embedding phase lines in the equilibrium value versus parameter graph.

In conclusion, Brady's group members were in a position that afforded them the ability to become legitimate peripheral participants in the summarizing of the changing structure of the solution space practice, which was fully introduced later in the class period through Lorenzo and Kenneth's presentation (see Part 3 below). Furthermore, Brady's group's lack of reasoning with dP/dt versus P graphs served as a constraint for them in adapting the new practice fully (see Part 4 below).

Part 3: Kenneth and Lorenzo's Presentation of the Fish.com Task

In this section we discuss Kenneth and Lorenzo's presentation, which lasted for approximately 17.5 minutes and culminated in the reinvention of a bifurcation diagram. Our analysis of this reinvention details all five forms of activity that constitute the changing structure of the solution space practice. In particular, we show how use of dP/dt versus P graphs was at the core of this group's analysis, which stands in striking contrast to the analysis we saw Brady present in Part 2. In addition, we document how the emergence of the summarizing the changing structure of the solution space practice was made possible through the brokering actions of the teacher and of Kenneth and Lorenzo.

Episode 3a: Introducing a New Inscription

Lorenzo and Kenneth began their report with Lorenzo recapping their analysis of the differential equation before it was modified to reflect harvesting. They use three graphs in their explanation (Figure 7.6).

> Lorenzo: This here, that's, uh, that's our initial equation without any harvesting, without anything. For that equation [points to $dP/dt = 2P(1 - P/25)$ on their overhead presentation] we can see that if we graph, uh, dP/dt versus P, uh, that's going to look like, that's going to be a parabola with two equilibrium solutions. So we can graph that too [points to the P versus t graph], and that's where we start from.

The starting place for their report is similar to Kurt's presentation detailed in Part 1, but less detailed. In particular, Lorenzo used a single dP/dt versus P graph to ascertain the structure of the solution space, which is aspect "a" (Figure 7.1) of the summarizing the changing structure of the solution space practice. As we have previously argued, this is in and of itself a separate CMP that becomes embedded in this emerging new practice.

Moving on to the next part of their presentation slide, Lorenzo explained how they algebraically dealt with the parameter in the modified differential equation and then how they graphically treated this algebraic result.

> Lorenzo: Now, we include this negative k [points to $dP/dt = 2P(1 - P/25 - k)$ on their overhead presentation] which tells us that we are going to harvest some fish. That k can be any number so we're just going to leave that as a constant. Now, we can solve this equation. We can set dP/dt equal to zero, like we did in this case [when $k = 0$], and try to find the two equilibrium solutions, but in this case, since we've got a constant, there are going to be lots of equilibrium solutions. Actually, we can graph those. That's the graph that we got right here [points to graph of P versus k, see Figure 7.7]. So this here, that's, uh, the graph of equilibrium solutions versus k.

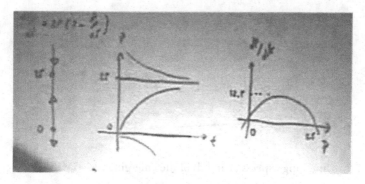

Figure 7.6. Analysis of initial differential equation.

In terms of the summarizing the changing structure of the solution space practice, Lorenzo just introduced to the class aspects "c" and "d" of the summarizing the changing structure of the solution space practice (Figure 7.1). This was the first time that these two specific ways of using the differential equation had appeared in class, and hence they represent Lorenzo and Kenneth's own creative efforts. Moreover, Lorenzo's introduction of these two forms of activity positioned him and Kenneth as brokers between their small group and the larger classroom community. As brokers, their presentation served the function of inserting new algebraic techniques and a new inscription, namely a "graph of equilibrium solutions versus k."

Lorenzo went on to explain how he and his group used the equations for P_1 and P_2 to interpret the number of equilibrium solutions as a function of the parameter k. This is aspect "c" of the emerging practice (Figure 7.1).

Lorenzo: Okay, from these two equations here [referring to $P_1 = 12.5 + 2.5\sqrt{25 - 2k}$ and $P_2 = 12.5 + 2.5\sqrt{25 - 2k}$, as shown in Figure 7.7] we can see that if you want to have an equilibrium solution, this root here has to exist, meaning this number here [pointing to the discriminant in the P_1 and P_2 equations] has to be either positive or zero. That means that our k cannot exceed 12.5. So if you want to have an equilibrium, you can't exceed 12.5, you can't take out more than 12.5 fishes per cycle: a year, a month, or whatever it is. Because if you don't have an equilibrium, that means basically all your population is going to drop to zero [moves his hand straight down in a way that seems to resemble a flow line in case when $k > 12.5$ [see Figure 7.8]. Now, if you, if k is zero—

Here we see Lorenzo carefully explaining how to interpret the equations for P_1 and P_2 in terms of the equilibrium solutions. The detail that he provided stands in contrast to the very brief explanation given for the

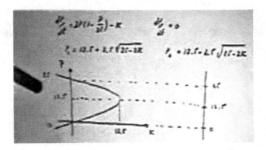

Figure 7.7.   Initiating aspects c) and d) of the emerging new CMP.

initial situation when $k = 0$ and the graphs (Figure 7.6). This suggests that Lorenzo recognized the newness of his and Kenneth's approach and so he took some care in explaining their idea.

Episode 3b: Making Connections

Before Lorenzo could continue, the teacher interrupted him to inquire about the case when there are no equilibrium solutions.

> Teacher: Hold on, hold on. I just want to interrupt for a second. Lorenzo, you just said that if you don't have an equilibrium solution your population is going to drop to zero? Is that what [Lorenzo: Yeah]. So that was similar to what you guys were concluding before, right? If you're taking out fifteen or twenty. So if your k is bigger than 12.5 it's over here [points to a k-value bigger than 12.5 on the k axis of the P versus k graph]. How do I understand what—I mean, there's nothing here [waves hand over the region to the right of the parabola on the P versus k graph] on the graph that Lorenzo is showing, right? Is there, is there something I can still understand from that graph that he's putting up there?

We see this interruption as a significant brokering move that functioned to provide opportunities for students to reflect on new ideas and inscriptions. Part of reflecting on this new inscription of P versus k was to mark similarities across presentations. For example, the teacher noted that Lorenzo and Kenneth's conclusion when k is bigger than 12.5, "was similar to what you guys were concluding before." After noting this similarity, the teacher went on to prompt students to make sense of the P versus k graph when k is bigger than 12.5. In particular, the teacher pointed to a place where "nothing's there" in Lorenzo and Kenneth's P versus k graph and asked, "is there something I can still understand." The significance of this brokering move is evident next when a student responded to the teacher's query by seeking to make a connection between different inscriptions, specifically between the P versus k graph when k is bigger than 12.5 and a dP/dt versus P graph.

> Dylan: What's that graph on the right there? Where the parabola's below P, down on the bottom right.
>
> Lorenzo: Yeah, that's uh, that's where I split this here [points to the dP/dt versus P graph on their overhead for the case when $k >12.5$. See Figure 7.8].
>
> Dylan: Because that is when the [equilibrium] solutions are imaginary.

Dylan's exchange with Lorenzo was the start of a more detailed conversation about how to make connections across the various inscriptions. As we see next, Lorenzo took the lead in explaining the mathematical details regarding the connections between the P versus k graph, graphs of dP/dt versus P, and the algebraic equations for P_1 and P_2. As such, Lorenzo was acting as a broker between his small group and the rest of the class in order to articulate the underlying mathematical ideas. Lorenzo began his explanation by revisiting connections for the "starting case" when k equals zero.

> Lorenzo: Yeah, I started, I wanted to include all the possible combinations that we can have for k. For example, if k is zero, that means you don't take any fish out, and then this radical here [points to the equations for P_1 and P_2] becomes five, and this, then you get two equilibrium solutions, twenty-five and zero. That's our starting case—that means we don't harvest anything.

It is significant that Lorenzo chose to tell the class that he "wanted to include all the possible combinations" for k. Some of the previous group

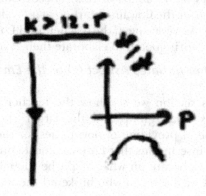

Figure 7.8. dP/dt versus P graph when $k > 12.5$.

presentations did not consider all possible cases for k. Thus, Lorenzo was providing a more thorough analysis than some of the previous presentations, and hence he may have found this worth noting. In addition, what constituted a new case for Lorenzo and Kenneth was when there is a change in the number of equilibrium solutions. That is, when there is a change in the structure of the solution space. Their P versus k graph offered the class a single graphical inscription that captured this change.

Next, Lorenzo explained some of the connections between the P versus k graph, the flow line, and the dP/dt versus P graph when $k = 12.5$.

> Lorenzo: If you take k to be exactly 12.5, which is this here [points to the vertex of the parabola in the graph of P versus k], then these two graphs of equilibrium solutions meet at this point here [see Figure 7.9a] and you have only one equilibrium solution [points to the place on the flow line when $k = 12.5$, see Figure 7.9(b)]. You can also see how if you graph dP/dt, that, uh, your population is going to be decreasing all the time except at one point [points to the vertex of the dP/dt versus P graph when $k = 12.5$, see Figure 7.9(c)]. That means basically the most fish you can take out is 12.5. That's the best case for you. But in case that your [initial] population is bigger than 12.5, because in that case you are going to come here and stop there [traces the flow line to the left of the dP/dt versus P graph shown in Figure 7.9b down to 12.5] and you are going to be taking 12.5 every time but your population still stays the same—12.5. If k is bigger than 12.5—

In the previous excerpt Lorenzo coordinated, in the case of $k = 12.5$, how to locate the one equilibrium solution on the P versus k graph, on the flow line, and on the graph of dP/dt versus P. He also interpreted this one equilibrium in terms of the fish.com scenario when he stated, "That means basically the most fish you can take out is 12.5." He then continued to explain that if the initial population was bigger than 12.5, then eventually the fish population would stabilize at 12.5. This explanation was accompanied by his tracing over the flow line, starting at a P-value bigger than 12.5 and ending at 12.5. As we see next, the teacher used Lorenzo's explanation of the flow line as a springboard to elaborate their P versus k graph.

Episode 3c: Teacher Initiates Aspect (e) of the Emerging New Practice

In the previous section we see how the teacher's brokering actions between Lorenzo's small group and the larger classroom community served the purpose of providing opportunities for the class to reflect on the new ideas and inscriptions that Lorenzo and Kenneth introduced. He did not, however, elaborate on what might be thought of as the mathematical content. It was Lorenzo who brokered the mathematical content by explicitly making connections between various inscriptions and equations. Lorenzo and Kenneth, however talented, were only able to go so far in their role as brokers of the mathematical content. It is the teacher who was positioned to help further develop the mathematical ideas initiated by Lorenzo and Kenneth in ways that were increasingly compatible with more conventional or formal approaches. In this way, the teacher acted as a broker between the classroom community and the broader mathematical community. In this particular case, the teacher stopped Lorenzo from continuing to explain how to see the various connections when k is bigger than 12.5 and began to elaborate Kenneth and Lorenzo's P versus k graph by layering a flow line on top of their graph.

Teacher: Hold on a second, I want to make sure that I understand. I was just think-ing that—can I draw on this, is that okay? [Lorenzo: Yeah.] So, um, you were saying that if k was equal to 12.5—this is your flow line with one equilibrium solution [points to the flow line for $k = 12.5$ shown in Figure 7.9(b)] [Lorenzo: Yeah.]. And your population is, you start with bigger than 12.5, it would be decreasing to 12.5. Any population below 12.5, if you're harvesting 12.5, would—you'd die out, right? [Kenneth: yeah, so you'd have to start] So it seems to me like right here [points to the vertex of the parabola on the graph of P versus k], I could get that information in a way. So when k is twelve point five, we have one equilibrium solution [draws a dot at the vertex of the parabola on the k versus k graph], and then the population, your P, are all decreasing here [draws a line segment with downward-pointing arrow above the dot just drawn]. [Lorenzo: Yeah.] And then over here it's all decreasing as well, right? [draws a line segment with down-ward pointing arrow below the dot]

In the previous excerpt the teacher took the flow line associated with $k = 12.5$ and he layered it on top of the P versus k graph. In terms of the summarizing the changing structure of the solution space practice, we reference this form of activity as aspect "e" practice (Figure 7.1). We see the teacher's layering of the flow line on top of the P versus k graph as a brokering move between the broader mathematical community and the classroom community because it was the teacher who knew that bifurcation diagrams contain more information than the graph that Lorenzo and

Figure 7.9. Making connections to the graph of P versus k when $k = 12.5$.

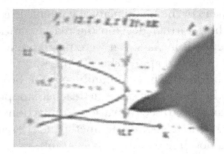

Figure 7.10.   Layering a flow line on the *P* versus *K* graph.

Kenneth created—bifurcation diagrams also contain information on the equilibrium solution type. By layering the flow line for $k = 12.5$ on the *P* versus *k* graph, the teacher was adding important mathematical information about the structure of the solution space when $k = 12.5$.

For Lorenzo, the teacher's actions were apparently not a major revelation because he immediately commented, "Sure, that's exactly what I've got here," pointing back to the flow line (Figure 7.9b). The teacher continued his brokering actions by next asking how to understand the *P* versus *k* graph in a similar way when $k = 0$. Again, Lorenzo responded in a way that suggests that the teacher's layering of the flow line was not surprising to him:

> The same graph, that's, see, here, I just showed you a few particular cases, but from this graph [the graph of *P* versus *k*] we can basically see all those cases, because this is exactly the graph of equilibrium solutions.

The teacher responded to this by continuing his brokering role. However, here he switched back to brokering between the ideas put forth by Lorenzo's group and the classroom community, as opposed to adding layers of flow lines, which served the purpose of bringing the mathematics of the local classroom community more in line with that of the broader mathematical community.

> Teacher: Hold on a second. So Lorenzo just said from that graph we can see all particular cases. [Kenneth: Yeah] Help us out. Kristen? I'm sitting in class. I'm not sure I understand what he just said. Help me out.
>
> Kristen: I'm not sure I understand. How could you see, like, what if there was fifteen? Like he was saying earlier

Perhaps not surprising given the complexity and interconnectedness between the many inscriptions, Kristen was uncertain how one can "see" all the cases as readily as Lorenzo apparently did. In the next Episode we detail how all the other cases can in fact be "seen" on the graph of P versus k.

Episode 3d: Teacher Hands Over Responsibility for Layering Flow Lines

Whereas it was the teacher who initiated the layering of flow lines on top of the P versus k graph, in this episode we detail how the teacher handed over responsibility of this task to Kenneth. While Kenneth had largely been quiet thus far in his group's presentation, here he stepped up and took over primary responsibility for discussing their report. To some extent, Kenneth's move to the center was prompted by the teacher, who, without Kenneth implicitly or explicitly requesting, gave the overhead marker to Kenneth and then stepped to the side of the room, putting Kenneth in charge. These subtle teacher actions fostered what Brousseau (1997) refers to as the devolution of responsibility. Rather than take on the question of how to interpret the case when $k > 12.5$, as suggested by Kristen, Kenneth chose to elaborate on the specific case of $k = 6$. He began his exposition by pointing to the equilibrium solutions on the P versus k graph corresponding to $k = 6$ (Figure 7.11): "You have a [harvesting rate] of six right here [points to where $k = 6$ on the P versus k graph], and you're going to have two equilibrium solutions right here and right here [draws in two dots on the P versus k graph corresponding to $k = 6$]." Kenneth's group had included graphs of dP/dt versus P, a flow line, and P versus t graphs this particular case at the bottom of their overhead presentation (Figure 7.12).

> Kenneth: So, like, down here, on the bottom [points to dP/dt versus P graph shown in Figure 7.12], you see that k equals six, our harvesting rate, right. So we have two equilibrium solutions. We graphed the differential equation against the population, so we get these two different equilibrium solutions here, right? [points to the two places that the graph of dP/dt versus P crosses the P axis]. So, if you were to graph P versus t for that, you'd see that you'd have an equilibrium solution here at 3.5 [points to the P versus t graph] and one here at 21.5. So you'd start out with a population, let's say a hundred, and then at that harvesting rate of six, you would just decrease to 21.5 over time. Similarly, if you were to start out with population anywhere between 3.5 or 21.5, you would just increase, or converge, to 21.5. So it's like, a sink, I guess you could say, would be our 21.5, and 3.5 would be our source. So if you were to start out anywhere underneath 3.5 for a beginning population with a harvesting rate of six, you would just decrease to zero over time.

Kenneth's explanation of the case when $k = 6$ address many of the features of what we referred to as aspect "a" of the emerging new practice. In particular, Kenneth details the nature of the two equilibrium solutions,

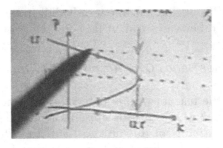

Figure 7.11.   The *P* versus k graph when $k = 6$.

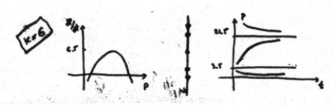

Figure 7.12.   The case when $k = 6$.

explaining why 21.5 would be a sink and why 3.5 would be a source. Next, the teacher prompted Kenneth to show the class how to reinterpret what he just said on the *P* versus *k* graph. Kenneth understood this request to mean that he should layer a flow line on top of this graph, similar to what the teacher did for the case when $k = 12.5$.

> Teacher: Alright, so show us this *k* equal to six up on this *P* versus *k* graph.
>
> Kenneth: So. This is six right here [writes in 6 at the appropriate place on the *k* axis of the *P* versus *k* graph as shown on Figure 7.13]. So this population here is 3.5. And this one here, what did we say it was, 21.5 [writes in these values on the vertical axis of the *P* versus *k* graph as shown on Figure 7.13]
>
> Teacher: So those would be my equilibrium solutions, you're saying, Kenneth?
>
> Kenneth: Correct. So this graph [points to the *P* versus *k* graph] just shows you for any different harvesting rate in between zero and twelve point five the various, uh, population equilibrium solutions.

The usefulness of Kenneth's explanation was demonstrated by Jim, who spontaneously added how to interpret the space inside and outside the parabola on the *P* versus *k* graph: "So in a way, the, um, inner part of that parabola is where the population is increasing [Kenneth: Correct.] and outside is where the population is decreasing." Kenneth then followed up on Jim's observation by "dropping" a phase line on top of the *P* versus *k* graph. Consistent with Jim's observation about the meaning of anywhere inside and outside the parabola, Kenneth then stated that you could drop a phase line "anywhere" on the graph, and chose to do so for $k = 6$. As such, the case when $k = 6$ represents a generic case for $0 < k < 12.5$.

> Kenneth: Right. Yeah, if you were to drop, like, a phase line on top of the graph anywhere, like if we were to drop one here [draws a vertical line at $k = 6$, see Figure 7.14] you'd see that increasing in there [draws an upward facing arrow on the line just drawn inside the parabola], decreasing there [draws a downward facing arrow on the line above the parabola], decreasing there [draws downward facing arrow on the line below the parabola].

In Part 4 students revisit how to determine the direction of the arrows on the flow/phase lines that are layered or dropped on a graph of P versus k. From Kenneth and Lorenzo's perspective, the way one decides on the direction of the arrows comes from analyzing graphs of dP/dt versus P. This is a form of activity that Brady does not engage in without the brokering by Kenneth, Lorenzo, and the teacher. As Kenneth continued his explanation, he returned to Kristen's question about how to interpret the P versus k graph in terms of the fish harvesting scenario when $k = 15$.

> Kenneth: So we can't harvest any more than twelve point five. You see, there's no population value associated with that [indicates the region to the right of the parabola]. You know, if you come out here to fifteen [writes in 15 on the k-axis and draws a vertical line at $k = 15$ without an arrow], there's no population out there, so,
>
> Teacher: Could you drop a flow line, a phase line there?
>
> Kenneth: You could drop them there, but they, I mean—
>
> Teacher: Would there be a flow line or phase line? [Several students speak up at once. Someone says, "It's always decreasing," and someone else says, "It would just be all gone."]
>
> Lorenzo: Of course, but it's just no equilibrium solutions.

As Kenneth, Lorenzo, and others in the class recognized, one could drop a phase line when $k = 15$, that is, at a k value where they are no equilibrium solutions. However, there seemed to be some reluctance to do so. This reluctance is understandable since it is the equilibrium solutions to which all other solutions converge or diverge. With no equilibrium solutions, there is less information to portray with a phase line. Importantly for the owners of fish.com, however, when $k = 15$ the solutions are "always decreasing" and hence the fish population is eventually "all gone," as students in the class pointed out.

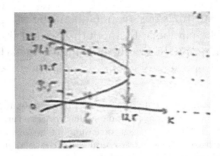

Figure 7.13.   Marking equilibrium solutions on the P versus k graph when $k = 6$.

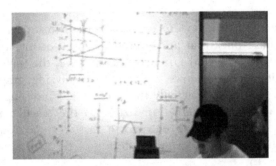

Figure 7.14.   Kenneth "drops" a phase line on the P versus k graph.

Episode 3e: Reasoning with a Continuously Changing dP/dt Versus P Graph

After a brief discussion about what harvesting rate students would recommend to the owners of fish.com and why, Roy asked Kenneth and Lorenzo, "How do you know where to put those arrows?" Roy's analysis had included a slope field and phase line based on that slope field, but did not include a graph of dP/dt versus P. Kenneth interpreted Roy's question as a request to re-explain how to use a graph of dP/dt versus P to ascertain the direction of the arrows on a phase line, and proceeded to elaborate on his earlier explanation when $k = 6$. Lorenzo, on the other hand, interpreted Roy's question in a different way. While Kenneth interpreted Roy's question in terms of how to draw a phase line from a single graph of dP/dt versus P, Lorenzo followed up Kenneth's explanation by explaining why the phase lines are essentially the same whenever k is between 0 and 12.5. He did so without prompting from the teacher or anyone else.

Lorenzo: Now, maybe for better explanation. You see this first equation [points to the equation $dP/dt = 2P(1 - P/25) - k$] over there? And, uh, you graph dP/dt versus, uh, versus P, yeah. See that's, the first equation, is, uh, quadratic function. If you include a constant, in the quadratic function, you just shift,

Kenneth: It shifts it up and down [Moves his hand up and down in a continuous manner].

Lorenzo: You just shift it down. [Kenneth: Yeah.] So basically, all this here, as far as k is between zero and twelve point five. These graphs [points to the graphs of dP/dt versus P on their overhead presentation] are going to look the same. It's only these equilibrium solutions, their place, is going to change, like, you know, this [the gap between the equilibrium solutions] can shrink a little bit [makes a pinching gesture over a graph of P versus t to dynamically show how the equilibrium solutions come together as k changes].

Lorenzo began his explanation with, "Now, maybe for a better explanation." We do not know what made this a better explanation for Lorenzo, but in our view, Lorenzo's explanation was "better" because he explicitly attended to a continuously changing dP/dt versus P graph in relation to the structure of the solution space. Kenneth also participated in this explanation by animating the change to the dP/dt versus P graph by raising and lowering his hand. In terms of the summarizing the changing structure of the solution space practice, Lorenzo and Kenneth's elaboration is captured in activity "b" of this emerging new practice. Essentially, Lorenzo explained how the structure of the solution space would remain the same for $0 < k < 12.5$, with the only difference being that the gap between the equilibrium solutions would "shrink a little bit."

In the five preceding episodes of this Part 3 we see many brokering moves by Kenneth and Lorenzo in terms explicitly inserting new mathematical ideas into classroom discussion. In this episode we similarly argue that Kenneth and Lorenzo continued such brokering between their small group and the larger classroom community by explicitly reasoning with a dynamically changing dP/dt versus P graph in order to infer changes to the structure of the solution space. In the next and final episode of Part 3 we see the teacher build on the mathematical ideas inserted by Lorenzo and Kenneth.

Episode 3f: Introducing Conventional Terminology

In this last episode of Part 3, the teacher began by rephrasing what Lorenzo and Kenneth just said, and then connected this recap to formal or conventional terminology. In this way, we see the teacher again acting as a broker, first between Kenneth and Lorenzo's small group and the larger classroom community through his rephrasing and then between the entire classroom community and the broader mathematical community by the insertion of formal or conventional terminology.

Teacher: Let me just say something now. I think that was really nice. I want to pick up on something that Lorenzo just said [Teacher moves from the side of the room to the front]. He said that if you have different k values between zero and, well almost, less than 12.5, he's saying that the flow lines are all basically the same. [Lorenzo: Yeah.] Essentially there's two equilibrium solutions, one of which is a sink and one of which is a source. So no matter whether your k harvesting rate is between zero and 12.5, not including 12.5, you're going to have two equilibrium solutions, like here [points to the P versus t graph for $k = 6$], one of which is a sink and one of which is a source so that basic structure [holds out two flat hand to indicate the sink and source equilibrium solution as show in Figure 7.15a] and the whole way in which the population versus time changes is similar. These [the equilibrium solutions] move together, is what you're saying, right? [Lorenzo nods] And then at this particular value of 12.5 for the harvesting rate something happens. Those two equilibrium solutions—

Kenneth: Become one.

Teacher: Become one [brings his hands together as shown in Figure 7.15c].

In Lorenzo's explanation about the change to the equilibrium solutions as k changes, he used a somewhat subtle pinching gesture to animate the changing structure. The teacher, in contrast, made a much more obvious gesture as shown in Figure 7.15. Based on the teacher's placement of his hands away from his body and over the graphs on the overhead, we suspect that the teacher's gesture was intentional, meant to be clarifying and informative for students, as opposed to a gesture that is spontaneous and idiosyncratic. As the teacher continued, he introduced the "technical" word for the value of the parameter when there is a change to the structure of the solution space.

Teacher: There's a technical word for that parameter value—It's called a bifurcation value. So this k is a parameter. This k's a parameter and this value right here [circles 12.5], the technical word for that, that occurrence, when your two equilibrium solutions change to one, is a fundamental change in the way in which all the flow lines look. They were saying basically all the flow lines in here look exactly the same, their equilibrium solutions are just pushed together a little bit. But when k is 12.5, there's a

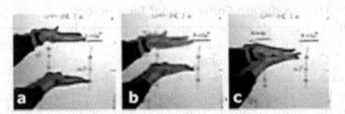

Figure 7.15.   Animating the changing structure of the solution space.

shift, or a change, in the way the structure of the flow line is. That parameter value is called a bifurcation value [Writes "bifurcation value" on the
overhead]. And then anything above 12.5, the flow-line, well we didn't
put it in here but from what you guys have said, the flow line here is
[draws in the missing arrow on the vertical line previously drawn by Kenneth on the P versus k graph when $k > 12.5$], there are no equilibrium
solutions, right? The population is just going to be dying out [points
downward along the flow line just drawn]. Right. So there's a flow line
here, it's just totally decreasing. [Lorenzo points to the flow line and dP/dt
versus P graph for $k > 12.5$.] Exactly, right, there it is. Right, so, this, that
parameter value here at 12.5, again, is called a bifurcation value, and it
just sort of helps us get a sense about when there's a big change, so here,

The class has now produced what an expert would recognize as a bifurcation diagram. Lorenzo and Kenneth first introduced what we might
now in retrospect call an "empty" bifurcation diagram, and through the
brokering actions of the Kenneth, Lorenzo, and the teacher, this empty
bifurcation diagram was elaborated on by dropping phase lines on top of
the graph. This elaboration was initiated by the teacher and furthered by
Kenneth, Lorenzo, and Jim. Finally, in contrast to more traditional teaching in which formal or conventional terminology is often the starting
place for students' mathematical work, this teacher chose to introduce the
formal mathematical language only after the underlying idea had essentially been reinvented by students.

In detailing the emergence of this bifurcation diagram, we articulate all
five forms of mathematical activity that constitute the summarizing the
changing structure of the solution space practice (Figure 7.1). In Part 4 we
see how all five forms of activity are relived in a new problem, one not
couched in a real world scenario, and how Brandon and his group
become more central participants in the new practice.

Part 4: Brady's Group Becomes More Central Participants in the New Practice

In this section, we detail Brady and Neil's presentation of their analysis
of a new differential equation, $dy/dt = ky + y^3$. Brady and Neil incorporated
important aspects of the previous solution of Kenneth and Lorenzo, but
also failed to use the important tool of the dy/dt versus y graph to interpret
the structure of the solution space. In the class discussion the teacher, and
at times Lorenzo and other students, pushed Brady and Neil to use this
tool to correct aspects of their solution and further interpret their results.
In Episode 4a, we describe the aspects of the summarizing the changing
structure of the solution space practice that Brady and Neil used in their
presentation and comment on which of these were new practices for their

group compared to their previous presentation. In Episodes 4b–4d, we describe how the new CMP continued to be disseminated to Brady and Neil and the classroom community through the brokering of Lorenzo, other students, and the teacher.

Episode 4a: Brady and Neil's Presentation

Brady explained what he and Neil had written on their overhead slide and answered clarifying questions from the teacher and other students. In this way we see Brady and Neil illustrating what aspects of the CMP that their group worked with and which they did not work with. Brady began with setting $dy/dt = ky + y^3$ equal to zero and solving to find $y = 0, y = \pm\sqrt{-k}$: "It made sense that when dy/dt is equal to zero you have an equilibrium point. So all we did was we set dy/dt equal to zero, and solve the equation to find out what y is equal to when dy/dt is zero. We got y equal to zero and y is equal to plus or minus square root of negative k."

In Part 2, Brady's group had presented a table with "POP MIN" and "POP MAX" values and had noted that these were equilibrium solutions. So it is likely that when Kenneth and Lorenzo initiated what we describe as aspect "c" of the new CMP (Figure 7.1) that this was a sensible strategy for Brady's group. Having implemented a similar strategy for specific values of the parameter on their own, it made sense to Brady's group to generalize this to an equation with the parameter as a variable as Lorenzo and Kenneth had done. Brady continued his explanation by describing how the graph on the overhead of y versus k illustrated these equilibrium values.

Figure 7.16.   Brady and Neil present their work on $dy/dt = ky + y^3$.

> Brady: Then we did the graph of all three of those functions and we got this here [points to y versus k graph with phase line drawn in]. And what we found was that as k got negative ... it decreased, then the gap between the two equilibrium points and the center equilibrium point became wider and wider. So we could make the statement that says, As k decreases the distance between E_1 and E_2 increases [Figure 7.17].

Here Brady illustrates aspect "d" (Figure 7.1) by describing how they have plotted the equilibrium values as they vary as a function of the parameter. This is not something that their group had done in Part 2. From the two previous excerpts we see that Brady and Neil had engaged in aspects "c" and "d" (Figure 7.1) in ways that they had not done earlier in the class period for the fish.com problem. However, there was no evidence on their overhead or in their discussion about using a dy/dt versus y graph, either with or without a parameter (aspects "b" and "a"). In Part 2 there was no evidence of Brady's group taking part in aspects "a" and "b" of the practice in their solution to the fish.com task. This lack of reasoning with dy/dt versus y graphs served as a constraint for them in adapting the new practice fully, as we will see in Episode 4b. The extent to which they have implemented aspect "e" is discussed in Episode 4b.

Episode 4b: Lorenzo and Others' Brokering Moves Regarding the Flow Lines

Lorenzo questioned the direction of the arrows that Brady and Neil drew on the graph of y versus k. Brady had not yet described these in his

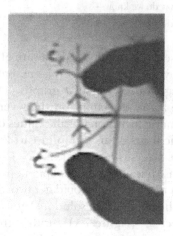

Figure 7.17.   Brady illustrates the distance between E_1 and E_2 increases as k decreases

presentation of the overhead, but the overhead included an attempt to put arrows on the y versus k graph in a way that mimicked Kenneth and Lorenzo's use of flow lines. Brady and Neil's arrows pointed up between the E_1 and E_2 curves and down when y was larger than E_1 or smaller than E_2 (Figure 7.17). This matched the direction of the arrows in the problem that Lorenzo and Kenneth illustrated in that they were up inside the curve and down outside of it (Figure 7.14).

Teacher: Lorenzo has a question for you.

Lorenzo: I think you guys got a problem with uh … (laughing in the room)

Teacher: Why do you all laugh? Don't laugh.

Brady: What's your, what's wrong with it?

Lorenzo: With the direction of those arrows on the flow line.

Brady: What's wrong with them?

Lorenzo: Well, you have two wrong directions. Okay, you've got four arrows. [Brady: Right.] How did you figure it out?

Brady: Oh, what is it … you're talking about these right here? [points to the y versus t graphs]

Kenneth: No. The, the phase line …

Jim: The way your arrows are going through your flow line.

Lorenzo: Two of them are right. Two of them are wrong.

Brady: Which two are right, you think?

Lorenzo: And uh, if you go from the bottom, that one is right. The next one (Brady points to the phase line) is right then. The other two are, yeah, should be the opposite way.

Teacher: How can we decide—

Kenneth: Yeah, how, how do we know—

Teacher: How can we decide?

From the above we see that Brady was not aware of the error and did not immediately have a way to think about this issue. Below we see that Brady responded by pointing to the y versus t graphs. The direction of those graphs did match the arrow directions on the y versus k graph. He also reverted back to a numeric focus for specific values of the parameter reminiscent of the numeric focus of his group in Part 2: "You can look at the function [points to the y versus t graphs].… No, I—it would, well—yeah, because it would increase because if you had put in negative nine you would get three. If you put in negative four, you'd get two. So it [points to the top curve of the graph of the equilibrium solutions] goes up." Here we see brokering regarding aspect "e" (Figure 7.1). Brady's group had imitated this aspect of the work done by Lorenzo and Kenneth's group. Also, they seemed to know that these arrows were representative of the direction the

y versus t functions moved, at least in some sense. However, they were not re-presenting the true structure of the solution space for the equation $dy/dt = ky + y^3$.

In this discussion, the line on the overhead graph of y versus k and its arrows served as a *boundary object* between the classroom community as a whole and the small group of Lorenzo, Kenneth and the teacher. Each group interpreted the boundary object differently. The small group, and perhaps some other students in the class, had come to understand a way to interpret this object with reference to the parameterized equation and the y versus k graph. Other students in the class, at least Brady, seemed to have a way to interpret the line as related to the y versus t graph but not with regard to the parameterized situation. Lorenzo and others pointing to this boundary object and questioning about the arrows served as acts of brokering between two communities about aspect "e" of the new CMP.

Episode 4c: Teacher and Others' Broker Moves Regarding the Use and Interpretation of a dy/dt versus y Graph

This episode began, immediately following Brady's comment above, with the teacher explicitly asking Brady and Neil to graph dy/dt versus y and interpret it: "Can you give us a graph of dy/dt versus y for k is equal to negative nine? That might help us decide whether or not the rate of change is positive or negative." After a couple moments of stumbling, Brady rewrote the equation for $k = -1$ and drew a correct dy/dt versus y graph for it. With the dy/dt versus y graph drawn, Lorenzo prompted, "Do you see what I'm saying now?" Brady seemed to immediately know how to interpret the graph of dy/dt versus y to say more about the flow lines and the solution space.

> Brady: Well, it's uh, it's increasing here [draws a plus sign over $-1 < y < 0$] and it's increasing there, right? [draws a plus sign over $y > 1$]. Negative there and negative there. So really, maybe we did mess up the top. So it really should go like that [switches the arrows at the top of the phase line on the y versus t graphs (see Figure 7.18)] and like that—
>
> Lorenzo: That's it.
>
> Brady: And then that, like that. So really then that messes up all this [points to the y versus t graphs].
>
> Kenneth: [softly] Yeah.

In this way Brady showed that he was able to competently engage in aspect "a" of the new CMP. This is not surprising in that aspect "a" was already itself an established practice in the class. However, this seems to

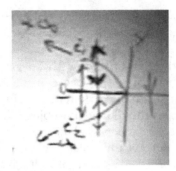

Figure 7.18.   Brady's corrected flow line.

be the first time that Brady had engaged in this practice in the context of discussing a parameterized situation. That is, we see here how the teacher calls for the use of the *dy/dt* versus *y* graph in a way that brokered between Brady and Neil's reasoning and that of Lorenzo and Kenneth. Here the *dy/dt* versus *y* graph served as a boundary object in that it was a familiar object for both groups, but one that had been interpreted and used by Lorenzo and Kenneth in their presentation in a way that Brady and Neil had not yet done. The teacher's act of calling on Brady and Neil to interpret a particular *dy/dt* versus *y* graph in this setting began the process of bringing Brady and Neil into more central participation in the new CMP by having them engage in part a) of the new CMP in this context.

Episode 4d. Teacher Broker Moves Regarding Interpreting Changing dy/dt Versus y Graphs

In this section, aspects "b," "d," and "e" of the new CMP begin to become more established in the class, including for Brady who, as the person at the overhead, functioned as representative of the students in the class who had not yet become a central participant in the new practice. Initially the teacher asked fairly direct questions to bring out a clarification of several important ideas and relationships involved in interpreting the *y* versus *k* graph.

Teacher: So how many, when *k* is negative, how many equilibrium solutions do you have?

Brady: When *k* is ne—we have three.

Teacher: Three. Does it matter whether *k* is negative ten or negative four or negative…. [Brady shakes his head says "Nope"] you've got three equilibrium solutions.

Teacher: What are they: sinks, sources, nodes?

Brady: The bottom one's a source. The second one is a uh ... sink. And the third one is a source.

Teacher: So I want to relate that back to something that Kenneth said, is that no matter what your negative k value was here. No matter what your k value is here, if it's negative, you get the same basic flow line. Just the distance be, distance between those equilibrium solutions is the same. Now that, in the previous problem, the bifurcation value, that special value of the harvesting rate, that parameter value was twelve point five. What's the, what's the special value of the parameter here, where there's a change in the kind of flow line you get?

Neil & Brady: Zero.

The above excerpt did not explicitly refer to the dy/dt versus y graphs, but it seems that Brady, having corrected the arrows using the dy/dt versus y graph for $k = -1$, then had a sense of what that means for all negative values of k. Thus the teacher's questions served to help Brady, and perhaps other students who were thinking like Brady, to move toward more central participation in aspect "b." Immediately following this excerpt the teacher engaged in another type of brokering by trying to engage other students in the class in the discussion. The teacher tried to get a student named Ted to take a more active role in the conversation, and then the teacher worked to get other members of the class to explain more about what a bifurcation value is and what it indicates about a solution space.

Teacher: Who else wants to do me a favor, just summarize the question that I asked. Dylan, yeah. Go ahead.

Dylan: Where is the bifurcation point?

Teacher: That's, that's clear and crisp, but say a little bit more about ...

Dylan: Where does the nature of your flow lines change? 'Cause on the left, you see there every flow line, every vertical line you take, um, is the same. Where does that change occur? Where the actual type of lines that you have will be different with the arrows.

Teacher: That's, that's a really nice way to say ... to somewhat elaborate on what does it mean to say what is that special value, that parameter, that bifurcation value where you have a change in the kind of flow line? [Pause] And so I heard, in the back here they said, "k is equal to zero." [Brady circles the point at which k is equal to zero]. Do you agree with that? Do you disagree with that as a, as the bifurcation value?

Brady: That's the point where all three graphs meet. Or where the three equilibrium solutions meet.

The teacher worked to help students better understand the formal term "bifurcation value." In this way he brokered between the mathematics community and the classroom community. The teacher and other students also continued to broker between those more central to the new practice and those more peripheral participants. To encourage a further elaboration of the nature of the changing space of solution functions, the teacher asked the students to explain more about the nature of the solution space when k is positive. This discussion began with further interpreting the equilibrium values in terms of the y versus k graph, aspect "d" of the CMP (Figure 7.1).

> Teacher: And what happens as k gets a little bit bigger, then?
>
> Brady: As k gets positive? [Teacher: Yeah.] Then you just have uh, dy/dt is uh— [Kenneth: Negative for all time.]—is negative.
>
> Teacher: So do you have any equilibrium solutions after that?
>
> Brady: No.
>
> Teacher: Kurt, say, say it out loud to the class.
>
> Kurt: Don't you have one still at y equals zero?
>
> Vincent: Yeah, at y equals zero we have one.
>
> Brady: Yeah, but I mean like as, as positive.
>
> Kurt: At positive values of k, you're still going to have y equals zero.
>
> Brady: Oh, that's right. [Lorenzo: Uh-huh.] I didn't even think about that.

Brady initially thought that the situation for positive k values would be the same as in the fish.com problem in which there was no equilibrium value and dP/dt was negative for all values of P. Here, with the help of other students, Brady realized that $y = 0$ is still an equilibrium value when k is positive. As the conversation continued, the teacher requested the class to think about the flow lines. Other students brokered the notion that the dy/dt versus y equations and graphs can still be used to determine the direction of the arrows. Although using dy/dt versus y graphs had been discussed before, Brady had not yet interpreted the dy/dt versus y graphs for the situation where k is positive and had difficulty immediately seeing how to do so. After four suggestions and comments from the teacher and eight comments from four different students, Kenneth described in more detail how to use the dy/dt versus y graph to interpret the flow lines for positive k values.

> Kenneth: We can, we can plug values into the, the dy/dt. Like plug zero in there. So we have just dy/dt equals y cubed. Which is uh similar to the graph on the bottom left. But there's no, um, little humps. It just uh, just kind of like, I mean you know what a y cubed looks like? So, we have increasing, when, uh ... when y is positive and decreasing when y is negative.

After Kenneth's comment, the teacher moved to the board and made a few clarifying points to broker between the notions proposed by Kenneth and other students who were trying to explain about using the dy/dt versus

y graph and those students, like Brady, who had not yet worked through interpreting the dy/dt versus y graph for positive values of k. The class ended with the teacher summarizing the nature of the equilibrium solutions and flow lines as k moved from being negative to zero to positive. In the next section, we distill the kinds of brokering moves that functioned as a mechanism for the initiation and emergence of the summarizing of the changing the structure of the solution space practice.

CATEGORIES OF BROKERING

In this chapter, we detail the emergence of the summarizing the structure of the solution space practice and the role of brokers in its initiation and ongoing evolution. This practice culminated in the collective creation of a complex and sophisticated inscription, namely that of a bifurcation diagram. Central to the emergence of this practice was brokering, which we demonstrated was a mechanism underlying the joint production of meaning. In this section, we step back from the fine-grained description of the emergence of the summarizing the structure of the solution space practice and reflect on what we see as more general brokering categories.

By definition, a broker is someone who has membership in more than one community, and brokering occurs when this person facilitates the infusion or appropriation of some form of activity from one community into another. The three communities that we identify are that of the broader mathematics community, the local classroom community, and the various small groups that comprise the local classroom community. The teacher in our data had the unique status of being the only person who was either a full or legitimate peripheral member of all three communities. Critical brokering moves, however, were not limited to the teacher. The undergraduate students in this class, who were full members of the local classroom community and their respective small groups, also carried out several noteworthy brokering moves.

Reflection on the various broker moves led to identification of the following three broad categories of broker moves: creating a boundary encounter, bringing participants to the periphery, and interpreting between communities. We follow Wenger (1998) in distinguishing between the terms boundary and periphery. The term boundary is more closely aligned with possible *discontinuities* between communities, whereas the term periphery is more closely aligned with possible *continuities* between communities. We hasten to point out that discontinuities and continuities are not dichotomies, but rather two ends of a continuum. Moreover, actual experiences of encounters between communities are simultaneously filled with a multitude of continuities and discontinuities.

We refer to the first general brokering move category as "creating a boundary encounter." A boundary encounter refers to direct encounters such as meetings, conversations, or visits between communities. Any boundary encounter will involve boundary objects. Boundary objects refer to objects that serve as an interface between different communities. A broker, by virtue of his or her membership in more than one community, is in a position to bring forth boundary objects that can facilitate encounters between communities. Wenger's examples of boundary encounters all entail direct meetings, conversations, or visits between communities. We adapt this notion to also include *indirect* encounters between communities. For example, except in rare cases, it is not feasible for direct encounters between the broader mathematics community and classroom community to occur. Instead, the teacher as broker can offer *indirect* opportunities for the classroom community to encounter the broader mathematical community. The teacher can do this by offering opportunities for students to engage in the broader discipline practices of modeling, symbolizing, defining, algorithmatizing, or proving. Such encounters allow for the possibility of participation in the authentic practice of mathematics, and hence provide occasions for the local classroom community to indirectly encounter the broader mathematics community.

We highlight two exemplary instances of this category of broker move from our analysis. The first example is between the local classroom community and broader mathematics community while the second example is between the classroom community and Lorenzo and Kenneth's small group. Our first example of creating a boundary encounter focuses on the teacher as broker in his role of selecting and constituting tasks. Because the teacher is a member of both the broader mathematics community and local classroom community, he is in a position to recognize characteristics of tasks that are likely to be productive for enabling newcomers to become more central participants in the discipline practices of mathematics, such as modeling and symbolizing. For example, the fish.com problem was recognized by the teacher as a task he could use to engage students in modeling by giving them an opportunity to revise the original differential equation to fit new assumptions. Modifying a differential to fit new assumptions was a novel task for students. In this case, the set of new assumptions and the original differential equation functioned as a boundary object because it allowed the classroom community to interface with the discipline practice of modeling.

After settling on a modified differential equation, the way in which the teacher constituted the remaining portion of the task was crucial in actually creating a further boundary encounter. In particular, the teacher constituted the task in such a way that it opened up the possibility for students to engage in symbolizing. Specifically, the teacher invited students to develop a report for the new owners of fish.com, rather than requesting students to recommend a harvesting rate. The later request would more likely lead to discussions about specific numerical values, rather than creating their own inscriptions to tell the story of what happens for all possible harvesting rate values. Constituting the task as a report to the new owners, without prescribing what form this report was to take, allowed students to develop their own inscriptions. Thus, the modified differential equation and task to develop a report functioned as a boundary object because it provided the classroom community an opportunity to encounter the mathematics community via participating in the discipline practice of symbolizing.

The second example of creating a boundary encounter comes from our analysis of Lorenzo and Kenneth's presentation of their work on the fish.com task. In inquiry-oriented classroom settings, it is fairly common for particular small groups to present their work on a problem to the entire classroom community. Such presentations represent the opportunity for a boundary encounter between one small group and their local classroom community. In our experience, not all such small group presentations realize this opportunity. For example, a small group presentation that simply reports back to the class what their group did without a substantive exchange of ideas and interpretations leaves the interface between communities somewhat empty. In order to fulfill the potential for a boundary encounter, there has to be some boundary object that leverages differences between the communities and actions by brokers to encourage an exchange of ideas and interpretations between communities. The P versus k graph presented by Lorenzo and Kenneth's small group was such a boundary object because this particular inscription was entirely new to the rest of the class and was the center of a substantive exchange between Lorenzo's group and the whole class.

In Episode 3a we highlight the detail and care in which Lorenzo explained his group's novel graph of P versus k. Our analysis describes how this presentation introduced two forms of activity that comprised the summarizing of the changing structure of the solution space practice, and how the introduction of these two forms of activity positioned him and Kenneth as brokers between their small group and the larger classroom community. Lorenzo and Kenneth's small group also included the teacher as a legitimate peripheral member. This is noteworthy because while Lorenzo's careful explanation of their P versus k graph opened the door for the classroom community to interface with their new inscription, it was the teacher who pushed the door open even further. As we detail in Episode 3b, the teacher interrupted Lorenzo and Kenneth's presentation to make explicit a connection to a previous small group presentation and to raise questions about how to interpret the novel inscription introduced by Lorenzo and Kenneth.

This move functioned to provide opportunities for students to reflect on the new forms of activity presented by Lorenzo and Kenneth. Thus, the first part of Lorenzo and Kenneth's presentation described in Episode 3a, together with the teacher's question about how to interpret their P versus k graph, created the opportunity for a boundary encounter between Lorenzo and Kenneth's small group and the classroom community. As our two examples of the more general brokering category of creating a boundary encounter illustrate, this particular type of brokering move sets up the opportunity and conditions for the boundary encounter to be realized. Our remaining two categories of brokering moves exemplify different ways that encounters between communities are realized.

The second general brokering move category we identified is "bringing participants to the periphery." Broker moves in this category help or encourage participants to move toward another community along a continuum. This is in contrast to the first category in which participants were set up to encounter the other community in a way some difference or discontinuity. In the preceding examples the broker's role was to create the opportunity for participants to engage a boundary object (task, inscription) that was relatively new to them. This set up a boundary encounter in the sense of being an encounter between discontinuous, distinct communities—those who would already know how to engage the task versus those who would not or those who had created and interpreted an inscription versus those who had never seen it before.

In comparison, "bringing participants to the periphery" is about moving along a continuum between communities. We tender two examples of this category of brokering move. Our first example of bringing participants to the periphery highlights the interplay between the local classroom community and the mathematical community. In Episodes 3c and 3d we see the teacher layering a flow line on top of the P versus k graph created by Lorenzo and Kenneth's group and then handing over responsibility for layering the flow lines to Kenneth. The initial layering of a flow line was a brokering move by the teacher that served more as the creation of a boundary encounter. This initial step introduced a new interpretation to the P versus k graph that brought it closer to the mathematical community's notion of a bifurcation diagram. The initial introduction of something completely new in this way is a brokering move that presents a discontinuity. However, as the teacher encouraged Kenneth to make further interpretations of phase lines for the P versus k graph this became more of a process of encouraging Kenneth, and the class with Kenneth as their representative, to reason and symbolize in ways that were more consistent with the broader mathematics community.

For our second example of bringing participants to the periphery highlights brokering moves between the local classroom community and smaller groups within this community. Parts 1 through 4 all include examples of the teacher acting as a broker to encourage members of one community to engage ideas of another community or for members of one community to explain their ideas more fully to another community. In this way the teacher is requesting that participants move toward another community through a continuous periphery. For example, in Episode 2b, after Brady has completed his initial explanation, the teacher said, "Brady, let's pretend that I'm an owner and I make a lot of money, and I'm an executive, but I'm not so sophisticated with tables and stuff. So I need some help understanding that table that you've got up there." This request for clarification asks Brady to extend the ideas of his group in more detail and more clearly to the rest of the classroom community. A few minutes later after further explanation from Brady, the teacher asks the class, "So that's a question to you all—does that explain what his table is? If you're the owners, are you understanding what he's, what information he's provided?" In this way the teacher requests the members of the classroom community to engage in the ideas of Brady's group and thus bring themselves toward the periphery between Brady's group and the classroom community.

The third general brokering move category that we identified is "interpreting between communities." As the label for this category implies, this particular category of broker move is one in which brokers facilitate the understanding of one community regarding how ideas are construed, notated, related, or labeled by another community. In comparison to the first brokering category, creating boundary encounters, this third type of brokering move occurs when a broker takes specific steps to fulfill or realize the opportunities that the creating boundary encounter moves offered. We provide two central, illustrative examples of this particular brokering category from our analysis. The first example involves brokering between the local classroom community and Lorenzo and Kenneth's small group. In the second example the brokering takes place between the local classroom community and the broader mathematical community.

Our first example of interpreting between communities comes from our analysis in Part 3. In Episode 3b, we see Lorenzo respond to Dylan's question about a particular dP/dt versus P graph (where $k > 12.5$). Lorenzo's response to Dylan's question went well beyond an answer to the particular question. As our analysis detailed, Lorenzo, acting as a representative for his small group, elaborated how the case where $k > 12.5$ fit with the other cases and how their group understood various connections to their novel P versus k graph. This explanation was significant because it served the purpose of framing how others' analyses could be interpreted in terms of one or more of these cases.

Lorenzo continued connecting his group's work to that of the others by explicating various relationships for the case when $k = 12.5$. Specifically, Lorenzo pointed to the vertex of the P versus k graph (Figure 7.9a) and explained that, "If you take k to be exactly 12.5 ... these two graphs of equilibrium solutions meet at this point here." Here Lorenzo made an explicit connection to their algebraic equations for $dP/dt = 0$ and their novel graph of P versus k. Lorenzo continued his explanation by purposefully linking the P versus k graph to two other inscriptions with which the class was more familiar. As he pointed to the node on the flow line (Figure 7.9b), he stated, "and you have only one equilibrium solution." He immediately connected this to a dP/dt versus P graph when he stated, "You can also see how if you graph dP/dt, that, uh, your population is going to be decreasing all the time except at one point," as he pointed to the vertex of the parabola (Figure 7.9c). Lorenzo was careful in his pointing, and through this care, he linked the inscriptions together by explaining how to construe the same information from three different graphical inscriptions. We see Lorenzo's pointing as a type of linking gesture that facilitated his efforts to interpret between communities. Linking gestures are often used to "provide conceptual correspondences between familiar and unfamiliar entities" (Nathan, 2008, p. 376). Here we see that Lorenzo was able to provide conceptual correspondences from what his classmates were already familiar with to a new, unfamiliar inscription through a careful use of pointing gestures. In relating what was familiar to the class to what was unfamiliar, Lorenzo's use of linking gestures facilitated his efforts to interpret between communities.

Our second example of interpreting between communities highlights the teacher's unique brokering role as the only person who is a member of both the broader mathematical community and the classroom community. Given this unique position, it is the teacher who can (re)interpret the mathematical ideas that are emerging in the local classroom community in terms of the conventional or formal terminology used by the broader mathematical community. In this way, the teacher can infuse formal terminology into the discourse of the classroom community.

In Episode 3f, we see the teacher setting up the infusion of the term bifurcation by using linking gestures to connect the familiar with the unfamiliar. One of the most familiar inscriptions for the classroom community was that of the P versus t graphs. The term bifurcation was unfamiliar to the classroom community, however the fact that the structure of the solution space is different for different k values was becoming increasingly familiar for students. In Figure 7.15, we see the teacher use a series of gestures that link the changing number of equilibrium solutions to the term bifurcation. In particular, the teacher extended his hands and forearms in a parallel manner to portray the parallel equilibrium solutions on the P

versus k graph, and then he brought his hands and forearms together (Figure 7.15c), at which point he explained that the "technical" term for the parameter value at which there is a change in the number of equilibrium solutions is "bifurcation value." Through these moves, the teacher explicitly introduced the conventional or formal term "bifurcation" at a point in the classroom discussion when it served the function of labeling an idea that was an emerging part of students' mathematical reality. We see this as a noteworthy departure from teaching that often starts lessons with formal or conventional terminology because it enables students to see themselves as capable of participating in the cultural practice of mathematics.

In this section, we introduce three generalized broker move categories: creating boundary encounters, bringing participants to the periphery, and interpreting between communities. Each of these brokering categories highlight the view that teaching and learning mathematics is a cultural practice, one that is mediated by and coordinated with the broader mathematics community, the local classroom community, and the small groups that comprise the classroom community. Because these categories were developed out of two days of classroom data, we make no claim that these categories are exhaustive. Furthermore, we contend that both the course content and the timing of the data observed influenced the categories' formulation. Thus, we expect that observing other data sets would result in the creation of additional broad categories or in the facilitation of a sharper definition of the existing categories through the creation of subcategories. It is to this end—observing more data sets for the expansion of the categories as well as for a sharpening of the existing categories—that we anticipate a direction for future research.

ACKNOWLEDGMENTS

This material is based upon work supported by the National Science Foundation under grant no. DRL 0634074. Any opinions, findings, and conclusions or recommendations expressed in this material are those of the authors and do not necessarily reflect the views of the National Science Foundation.

NOTE

1. An autonomous differential equation is one that is of the form $dy/dt = f(y)$. In other words, the rate of change in some quantity y depends explicitly on the quantity y only. Thus, $dy/dt = 3y$ is autonomous, but $dy/dt = 3y + t$, for example, is not autonomous.

REFERENCES

Brousseau, G. (1997). *Theory of didactical situations in mathematics*. Dordrecht, The Netherlands: Kluwer Academic Publisher.

Cobb, P., & Yackel, E. (1996). Constructivist, emergent, and sociocultural perspectives in the context of developmental research. *Educational Psychologist, 31*, 175–190.

Nathan, M. J. (2008). An embodied cognition perspective on symbols, grounding, and instructional gesture. In DeVega, M., Glenberg, A, M. & Graesser, A. C. (Eds.), *Symbols, embodiment and meaning: A debate* (pp. 375–396). Oxford, England: Oxford University Press.

Rasmussen, C., & Stephan, M. (2008). A methodology for documenting collective activity. In A. E. Kelly, R. A. Lesh, & J. Y. Baek (Eds.), *Handbook of innovative design research in science, technology, engineering, mathematics (STEM) education* (pp. 195–215). New York: Taylor and Francis.

Rasmussen, C., Zandieh, M., King, K., & Teppo, A. (2005). Advancing mathematical activity: A view of advanced mathematical thinking. *Mathematical Thinking and Learning, 7*, 51–73.

Roth, W. -M., & McGinn, M. K. (1998). Inscriptions: toward a theory of representing as social practice. *Review of Educational Research, 68*, 35–59.

Stephan, M., Bowers, J., Cobb, P., & Gravemeijer, K. (Eds.). (2003). Supporting students' development of measuring concepts: Analyzing students' learning in social context. *Journal for Research in Mathematics Education, Monograph No. 12*. Reston, VA: National Council of Teachers of Mathematics.

Stephan, M., & Rasmussen, C. (2002). Classroom mathematical practices in differential equations. *Journal of Mathematical Behavior, 21*, 459–490.

Wenger, E. (1998). *Communities of practice: Learning, meaning, and identity*. Cambridge, England: Cambridge University Press.

CHAPTER 8

INSCRIPTION, NARRATION AND DIAGRAMMATICALLY BASED ARGUMENTATION

Narrative Accounting Practices in Primary School Mathematics Lessons

Götz Krummheuer

We *see* cause. (Bruner, 1986, p. 17)

My interest lies in the everyday situation of the primary school mathematics lesson. Of particular interest to me is how mathematical learning is made possible during social interaction. My research is hereby empirically delivered: my colleagues and I[1] observe mathematics lessons and videotape normal as well as experimental lessons. Using interpretive techniques, we reconstruct the way in which these regular or experimental lessons develop through interaction.

The perspective adopted here is a "situational" one: "My perspective is *situational*, meaning here a concern for what one individual can be alive to at a particular moment, this often involving a few other particular indi-

Mathematical Representation at the Interface of Body and Culture, pp. 219–243
Copyright © 2009 by Information Age Publishing

viduals and not necessarily restricted to the mutually monitored arena of a face-to-face gathering" (Goffman, 1974, p. 8). The concept of the "situation" refers here to a human social gathering, thus meaning a social situation like a "social event" or "social occasion." It is characterized by this feature: individuals interact with one another and that by this a situation as a social one emerges. The focus is directed towards the "here and now" of a situation, on the exchange and matching of actions between those present, thereby creating the possibility of joint actions.

With relation to this perspective, our empirical studies focus on the way in which, with respect to the everyday conditions of the school lesson, the learning of mathematics is enabled and supported but at times also hindered. The posing of such a stance is above all indicated as learning in the sense of a mental activity arises fundamentally from the participation in such social situations:

> human mental activity is neither solo nor conducted unassisted, even when it goes on "inside the head."... Mental life is lived with others, is shaped to be communicated, and unfolds with the aid of cultural codes, traditions, and the like. (Bruner, 1996, p. XI)

We recognize learning, and here in particular mathematical learning in the primary school, to be a social process, which derives from the interaction between those taking part in the lesson, in other words it is the result which arises from this interaction. The mathematics learning process stems from a coordination of several individuals' actions and corresponding mental activities. Usually this process is regarded as the "negotiation of meaning." Our interest is in the theoretical reflection and empirical analysis of this type of discourse-based learning process in the primary school mathematics lesson. During the negotiation process, the participants simultaneously create a rational basis for their negotiations.

The thesis of the social constitution of learning is inevitably linked to the idea that in principle it is only possible to learn through indirect means. Learning is enabled, facilitated and directed but never automatically caused through the participation in appropriate social interactions processes. Using these perspectives Birgit Brandt and I have constructed a model of the everyday mathematics lesson. It encompasses structural dimensions, whose situational realizations allow describing various degrees of optimizing the social constituents. These dimensions are the (a) thematization of a subject-based content ("how does the mathematical theme develop"); (b) accounting practice ("how is it established and explained"); (c) structure of interaction ("when is the student's turn"); (d) form of participation for actively involved students ("how can students actively take part in the lessons"); and (e) form of participation for non-actively involved students ("what happens to the quiet students?"). These five dimensions present fundamental aspects of the process of interaction during the everyday lesson, with regard to the conditions of the possibility of learning mathematics. In recent years ongoing work has been carried out looking at the problem of the participation of students in such interactive processes, particularly within the framework of this approach. It dealt with the interplay of the accounting practice and the forms of participation. In this chapter, the interdependence of the thematic development and the accounting practice are explored in greater depth. Hereby I include both the verbal expressions and the written contributions of the participants working through a given task. In particular the inclusion of the written notes in the reconstruction of the negotiation process leads to a systematic integration of a semiotic approach. I expose that the accounting practice along with the thematic development is narratively structured and that the written elaborations in this narrative structuring fit in as visible "steps" in the process of the interactively created argumentation.

I begin with an example in the next section, there after I take an in-depth look at the dimensions of the theoretical development and the approaches expanded within them and finally I will deal with at the accounting practices. The results of the analysis are summarized and commented upon from a narrative, theoretical perspective. I end with methodological comments about the analyses explored in this chapter.

AN EXAMPLE

The episode comes from a first-grade class and was designed to help students move beyond adding single-digit numbers. During a teacher-led phase of the lesson, several decompositions of numbers between 11 and 20 were demonstrated with attention to arithmetical sequences. Using diagrams, several decompositions of numbers had been worked out on the board. The following episode derives from a phase during which the pupils first had to work through a worksheet individually. Pupils who quickly finished the task were asked to help other pupils who were still working on the sheet. The discussion refers to the section of a worksheet (Figure 8.1). The circles on the original were shown in two colors—yellow and red. I refer to the boxes on the right-hand side, which are divided into three sections, as a whole as "blocks," and the sections as upper, left-hand and right-hand boxes respectively.

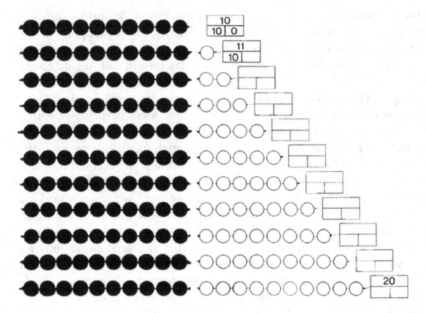

Figure 8.1.   Worksheet.

Wayne has been asked by the teacher to be a helper to the other children, and show them how to do the task: how you should work it out and what to write in the gaps. He takes his worksheet and turns to Efrem. The following dialogue evolves:

429	Wayne	*comes with his worksheet and places himself next to the smiling Efrem*
430		this is ten\ *points to the box on Efrem's sheet* and this is one. there you
431		have to / .look\ *(inaudible)*
432	Efrem	yes \ one \ yes
433	Wayne	yes + that's got to be one (ok put it in) *(inaudible)*
434	Efrem	And then *still pointing to the row of yellow circles from left to right*
434.1		two three four five
435	Wayne	*points at the box* ok look /
436	< Efrem	Six seven eight
437	< Wayne	*(inaudible)*
438	Efrem	nine ten. right here we've got to put in one + \ ok /
439	Wayne	*points to the box* look and here you've just got to put in a one
440	Efrem	*gets a pen out of his bag and writes one /*
441	Wayne	*points to the box* here at the top you've got to put in twelve /
442	Efrem	you mean another *(inaudible)*
443	Wayne	no \ twelve \ .. good / *points to the box and then here / . looks at his*

429	Wayne	*comes with his worksheet and places himself next to the smiling Efrem*
444		*worksheet*
445	Efrem	ten / *writes*
446	Wayne	*looks briefly at his sheet* uhhuh / ten \ .. and then you've got to
447	< Wayne	*(inaudible)* *turns to Carola (inaudible)*
447.1	< Carola	*turns her head towards Wayne (inaudible)*
448	Efrem	And two /
450	< Wayne	*Nods* mmm / can you do it now /
451	< Efrem	writes erm \
452	Wayne	look \ that's *pointing at the box* twelve thirteen fourteen
453		Fifteen sixteen seventeen + and that / in this box that's always ten *(inaudible)*
454	< Wayne	yes \ yes \ taps on the table with his pen
456		so \
457	*Efrem continues to work on his own.*

Already at the start Wayne is probably referring to the spot where there is something to be filled in (<430>). After this he concentrates most likely on the numbers themselves that need to be written in. Thereby he progresses block by block from <432, 439> to the third block <441, 446, 450>. By doing this, he takes as his theme the numerically and diagrammatically related segmentation of the left- and right-hand digits. In order to do this, he does not need any conceptual idea of tens and ones in terms of the decimal place value system. In contrast it appears to me that Efrem's contributions at the beginning of the episode show that he is focusing on the yellow (white) circles. He refers to a continual increase of one yellow circle in each row going from the top to the bottom <434, 434.1,436> He wants, if possible, to record this constant increase in the right-hand box of the second to the eleventh rows, by writing a one in each of these boxes <438>. Efrem grasps, by focusing on the pictorial representation, that the nature of the task involves a vertical system of the yellow circles increasing by one each row.

His approach does not find acceptance in the interaction with Wayne: the yellow circles play less and less of a role. At the end Efrem only refers (from <440>) to the diagram of the block. But his synoptic view of the complete diagram and his clearly perception of the vertical system remains given. Wayne accepts this system in <452> and Efrem reinforces it in <454>. The *connection* brought up by Efrem between the representation of the circles and blocks gets lost again however during their cooperation. Wayne explains the system of the worksheet by predicting the order of the numbers "10 and 11" using their canonical rising as positive integers. He takes these two starting numbers from the uppermost boxes of the first and second rows. Wayne withdraws from the helper position after he has been convinced of the "success" of his help <450, 455>. Efrem completes his worksheet following the reproduced episode by himself.

THEMATIC DEVELOPMENT AND INSCRIPTION

This section examines two aspects: I describe the fundamental concepts of the negotiation process of meaning and subsequently explore the function of written notes and drawings within such a negotiation process. First, in the situational perspective chosen here, during a particular mathematics lesson, a negotiation takes place to ascribe a specific meaning to the themed mathematical terms. Word by word the participants discuss and agree on what, to their minds, can to be seen at this moment in time as mutually shared meaning. Three comments may help clarify this concept further. This type of discussion does not necessarily have to take place in detail or be definitively explicit. In the above example the different views of the two students become thematic. They are not however expressed explicitly and so for Wayne and Efrem are not comprehensible on a conceptual level.

The participants will often interpret the situation in ways that are familiar to them, and not necessarily continually create new interpretations for themselves:

> Presumably, a "definition of the situation" is almost always to be found, but those who are in the situation ordinarily do not create this definition, even though their society often can be said to do so; ordinarily, all they do is to assess correctly what the situation ought to be for them and they act accordingly. (Goffman, 1974, p. 1)

With reference to Goffman I speak here of a "framing." In the given example we must understand Efrem's first interpretation in the sense of a vertical system not as an original and unique interpretation. It can equally well be based upon the "experienced" view of a student who is searching for a rule between the graphic and the "number" blocks. I discuss this point further below.

One can ascribe to Efrem a "framing" whose theme is the relationship between the diagram of yellow circles and the numbers (to be filled in) inside the blocks. It comes to him when he identifies a systematic line-by-line increase of yellow circles in the graphic. I call this the "graphic-number-framing" and in short the "graphic-framing." Wayne however finds the numbers for the blocks by decomposing the two-digit numbers into tens and ones on a graphic level and then filling them into their appropriate boxes. I put him in control of a "digit-number-framing" (or "digit-framing"). As illustrated above, this framing does not necessarily consist of a developed concept of a number based on a place value system.

The term negotiation of meaning may give the impression of an ideal symmetry between speakers: equal partners searching for a communal interpretation. In reality that is of course rarely the case. Even in asymmetric interactive processes, negotiation of meaning take place. In the above example, because of Wayne's helper role one can presuppose an asymmetric interaction. In the analysis of the episode Wayne can demonstrate a certain level of "superiority." During the course of the dialogue however, a "working consensus" emerges that is not only shaped by Wayne's ideas, but also takes in, by way of emphasis on the vertical system, Efrem's way of looking at the problem. Using the term "production format" we have looked in some detail into the phenomenon of "participating contribution" described in the thematic development in the so-called fourth dimension from the first section (forms of participation for actively involved students).

Second, it is typical for mathematics conversations to include integral written graphic elements within the language of everyday speech. In the example described in the above text, this phenomenon is evident in the presentation provided by the worksheet including a diagram as well as the "graphic" presentation of the two-figure left- and right-hand numbers. The facts discussed here expand, however, the concrete nature of the task: mathematical communication is hardly imaginable without the consultation of a written-graphic element. To describe these written-graphic recordings the term "inscription" is used here. I understand the term to refer to "pieces of craftwork, constructed in the interest of making things visible for material, rhetorical, institutional, and political purpose. The things made visible in this manner can be registered, talked about and manipulated" (Roth & Mc Ginn, 1998, p. 54).

Inscriptions can be the conventional representations of numbers in the decimal system, didactically motivated illustrations, individually made up drawings and much more. In mathematics education rather frequently adopted concepts like "note taking," "mathematics journals," and so on do not adequately cover the interesting interim-situational aspect of such written recordings referred to here. Inscriptions are introduced, made up, situationally adapted and further developed during the negotiation processes. In more recent works on semiotics in mathematics education this viewpoint is considered by Charles S. Peirce's concept of the "diagram" or "diagrammatics." Diagrams are inscriptions that result from a "(conventional)" system of rules governing production, use and transformation" (Dörfler, 2006, p. 202; translation by GK). In particular Peirce emphasizes the assumed adherence to the rules for altering and further developing such inscriptions.

These kinds of notes, sketches, drawings, and so forth, are of interest in view of the situational perspective considered here, above all when they are included in an oral-vocal led interactions process for the purpose of demonstration, clarification and support. The making up of inscriptions in interactive mathematical subject development already takes place in pre-school, when for example children are coloring in shapes, writing numbers for the first time or are completing pictures of simple figurative number sequences in interactive situations. An interactions process for mathematical thematic development contains therefore an oral-vocal dimension and in addition usually a visual-manual dimension.

Inscriptions are interesting in the *situational* perspective chosen here especially when they do not yet exist as a finished text simply adopted communally during the interaction, but when they still occur as a result of the process of *negotiating meaning*. In this respect the analysis of the above example points less towards the inscriptional elements of the worksheet. What is far more interesting is the genesis of new inscriptions, like, for example in the above episode, during the collective tackling of the problem, the graphic part of the problem is further worked through and thereby rules for the admissible manipulation of the inscriptions are employed, determined, or modified. The focus is directed towards whether and how a *diagram* occurs in the sense described above. Thus, in the above classroom episode the following inscriptions are developed: Efrem at the start enters a one in the right-hand box of the second row (Figure 8.2a). Following this Efrem fills in the block in the third row with the numbers twelve, ten and two (Figure 8.2b). Following the episode Efrem finishes the worksheet. This section of the worksheet is thereby completed. A further, more extensive inscription results from the work carried out on the task. Later, it can also prospectively be used for open class discussion. Bruner (1996) calls inscriptions in such cases "Oeuvre."

Often, however, these inscriptions are not designed for a broad readership but serve only the current negotiation process or as notes for one's own individual attempt at finding a solution. During an interaction inscriptions can vary in comparison to their medial dimension of the visual-manual (phonic versus graphic) and also in comparison to a conceptual dimension between communicative immediacy and communicative distance (Fetzer, 2003). The communication between a textbook writer and students is graphically clearly defined from the point of view of the medium and is conceptually molded by a great communicative distance. An inscription created spontaneously during interaction between students working together is characterized by a great communicative proximity. If in the ideal case scenario the schoolbook is to be understood without any direct interaction with the author, then inscriptions that are

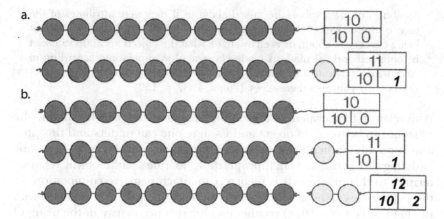

Figure 8.2.   Efrem's moves.

spontaneously accomplished among students cannot claim to be typically conceptualized for such a communicative distance. It is possible that just a short while later the participating students would no longer be able to interpret their own inscriptions appropriately.

In the example in this essay the self-developing inscription changes its conceptual positioning. If one can initially give the inscription a close communicative proximity between the students, then by the end of the task the communicative distance increases: at the end Wayne is no longer present. The final worksheet does not then in any way represent the common solution process. As a reader of this work one could think that the students have solved the task by activating a developed understanding of the decimal place value system. This is however—as the reconstruction shows—not the case. The situationally relevant meaning of the developing inscription is subject to the interactive negotiation process and is thereby a component part of the thematic development.

In many such negotiations the *representation* of an object is taken in the form of a sign rather than the mathematical idea or procedure itself and should not exclusively be seen as its representation. In this way Efrem for example formulates in <432> a one as he refers to the yellow circles in the second row. The yellow circle is identified with the number 1. In the same way Wayne handles the number 12 in accordance with the possible manipulation relating to the symbol "12."

One can conceive as problematic particularly Wayne's dealings with the sign "12" from a normative-mathematical didactical viewpoint. Such an identification of a sign or symbol with a mathematical term or object is a typical procedure for mathematical inscriptions:

> Attributes of the symbols are talked about as if they were attributes of the objects themselves.... It encourages one to follow a calculation without being concerned about, or even aware of, what the objects are.... Results can be obtained and calculations made via symbol manipulation, according to rules which relate to the symbols rather than dealing directly with operations on the concepts themselves. (Pimm, 1987, p. 159)

With relation to prevalent dyadic semiotic approaches, which examine the relationship between an object and its sign, one can understand this phenomenon of mathematical usage of signs most likely as a "semantic pathology." I consider it inappropriate to ascribe "pathological" characteristics to this practice. It is incorrect to attach negative connotations to the way we view signs as often useful but not unhelpful. Alternatively with the help of Peirce's (1978) triadic semiotic this peculiarity in the using of mathematical signs explains and characterizes itself as specificity in the sense of a standardization or subject-cultural framing.

Peirce's (1978) approach cannot be outlined in greater detail here. We will only look into the differentiation between the terms "representamen" and "interpretant" of his treatise. To describe a sign that stands for an object, Peirce uses the term representamen. The individual interprets this object–representamen relationship. The product of the interpretation is called the interpretant. Categorically seen this is again a sign. Its object is now a dual relation that consists of the original object and its given representamen. Principally the representamen does not stand for all aspects of its object. It is much more a specific characterization of the object that is carried out through this sign. Peirce talks here about the "ground of the representamen."

The peculiarity of the identification of the sign (representamen) with its object in mathematical inscriptions as described above, which is for example illustrated in Wayne and Efrem's framings at the beginning of the episode, can be understood with the help of Peirce's approach of a specific form of the interpretant in such a triad: the object is identified with its sign. Often the manipulation of the *signs* makes possible a new understanding of the qualities of the corresponding *objects*. In particular the works on diagrammatics and the associated reglemented practice of manipulation of inscriptions clarifies that these interpretants are not just spontaneous products of interpretation by single individuals. It is far more often a case of a conventional or sometimes standardized handling of these diagrams, which is based upon a commonly accepted framing of situations using such diagrams.

I ascribe to Efrem a graphic-framing of the task situation when he starts working on the problem. As regards Wayne I assume that he interprets the task situation in terms of a number framing. In the course of their negotiation of meaning a new framing appears, which one could call a *counting-number-framing* or for short "counting-framing." The numbers from 11 to 20 are decomposed into tens and ones and then the right hand box is filled in counting from top to bottom with a ten plus a zero and then from top to bottom with the ones.

During the negotiation process a new framing has emerged. This one clearly is shared between the two students. Furthermore, the framing process depends fundamentally on the system of inscriptions as the students interpret them. Within this framing the inscriptions become diagrams in Peirce's sense, which can be manipulated according to framing specific rules. The range of application of such framings is located within the spectrum between communicative proximity and communicative distance. If one can relate the initial framings of Efrem and Wayne even more strongly to the pole of communicative proximity then the counting-framing at the end of the working-through process is shifted more towards the pole of communicative distance.

ARGUMENTATION AND ACCOUNTING PRACTICE

In this section I initially introduce the term "accounting practice" and a corresponding analytical process. Using this process I reconstruct the various lines of argument from the above episode.

Terminology and Analytical Processes

In everyday interaction we assume that it is normally clearly indicated how seriously or rationally a concrete act should be regarded, through the action itself. Actions can be fun, serious, superficial, deeply thought out, spontaneous, or well-prepared. In many cases we regard this action without any complex additional interpretive help from the person performing it: "the activities whereby members produce and manage settings of organized everyday affairs are identical with members' procedures for making those settings 'account-able' " (Garfinkel, 1967, p. 1). I differentiate between the *performance* of actions and their *accountability*. Should any doubt arise about the presumed rationality of the action, then the participants can change to another form of discourse and debate this point explicitly. The mode of interaction changes because something is explicitly claimed to be disputable. This mode is "rational discourse" as opposed to the indisputable "communicative action" (Habermas, 1985).

The specific characteristic of the relationship between the performance of an action and its accountability is called an "accountings practice" (Garfinkel, 1967, p. 280). Within it, emerge performance and accountability separately ("discursively") or performance and accountability emerge coalesced in the flow of the interaction ("reflexively"). Accordingly I talk about a "discursive accounting practice" or a "reflexive accounting practice." Although initially in the above analysis of the example, different framings of Efrem and Wayne are reconstructed, no explicitly resolved disputability emerges. The two students realize a reflexive accounting practice, which is successful in terms of the fact that, at the end, they both solve the task conjointly within a shared counting-framing.

Within the accomplished accounting practice, one can attempt to reconstruct the argumentative aspect more closely. Reasoned argument is not only understood here as the characteristic of a discourse specifically aimed at producing a consensus. In many situations the arguments are produced at the same time as the actions in accordance with the above quotation from Garfinkel. One can understand Wayne's actions in the given example in the sense of a reflexive accounting practice: he recites up to <444> the answers he has already written on his worksheet and includes Efrem in his explanation, giving him clear instructions as to how he should look closely at his sheet. Efrem appears to be able to enter into Wayne's implicit argumentation and leads in <448> Wayne's argumentation to its conclusion in that he independently finds the missing number for the right-hand box in the second line.

Toulmin's (1969) functional argumentation analysis is applied for the reconstruction of such accounting practices with underlying argumentation. A person's spoken contributions are thereby viewed independently from their individual intention and interpretation as regards their function in a jointly created line of argument. The four central categories of an argumentation in Toulmin's terms are the "data," the "conclusion," the "warrant," and the "backing." The conclusion is a statement that should be substantiated. The data is an undisputed fact or a piece of information to which one can refer to as an answer to the question "What do you see as given?" The shortest possible argumentation would therefore be: data, so conclusion. It is reproduced in the uppermost line of the layout illustrated below. We call this line also the "inference." A kind of "accessible" conclusion can then be referred to later on as new data. Warrants are general, hypothetical statements, which can serve as "bridges" and legitimize these kinds of inferences. Generally speaking they correspond to a further possibility of argumentation and can be thought of as an answer to the question "How do you get there?" Backings are finally convictions that lead to the applicability of a warrant. They answer the question "Why should a particular warrant be accepted as generally reliable?"

Alongside the four parts making up an argumentation, Toulmin introduces additional categories that are not taken up here. In the episode, several such arguments make up a "multiple" line of argument. Here a conclusion functions simultaneously as data for a subsequent argument. Figure 8.3 shows the pattern that arises for such multiple argumentations. The well-known "Toulmin layout" (Toulmin, 1969) arises if one restricts oneself to the sections of the diagram that contain a "1."

Argumentation in the Example of Efrem and Wayne

In the episode one can identify each reconstructed framing with its respective argumentation. Consequently a subdivision of this excerpt presents itself in accordance with these framings (Wayne's digit-framing, Efrem's graphic-framing, and at the end the counting- framing shared by both of them).

Argumentation for the Digit-Framing

Wayne's actions produce suggestions (answers) for the numbers to be filled into the boxes on the second and third rows. These are the number 1 for the right-hand box on the second row and the numbers 12 (upper box), 10 (left-hand box) and 2 (right-hand box) in the block of the third row. In the sense of a reflexive accounting practice a two-fold argumentation emerges. With "this is ten and this is one" <430–431> Wayne cites the first item of data. It is shown here that it concerns the number ten which is already filled in, in the block in the second line. It is less clear to which one Wayne is referring. By "looking," it becomes obvious to Efrem through Wayne's suggestions that a one should be filled in, in the right-hand box of the second line <433, 439, 440> (first warrant). With the inscriptional fixation of the one, Efrem accepts both the concluding possibility, which has only been touched upon by looking, as well as the result. The request to look is legitimized by the fact that Wayne is in an official helper role and moreover has the "right" answer in his worksheet and pointedly refers to this. The conclusion "1" in the right-hand box in the

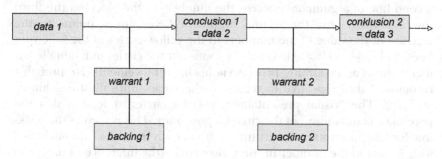

Figure 8.3.   Pattern of multiple argumentations.

second row functions in the further course of the interaction as a second piece of data for a subsequent argumentation. Due to Wayne's role as helper, the answers on his worksheet can be viewed with greater validity. In this sense his *correctly answered* worksheet acts as backing.

The second inference is based upon the block now filled in, in the second row. With the twelve <441, 443> Wayne initially refers back to the rising number sequence in the first two upper boxes. At the root of this one can see the system of the canonical rising number sequence of the positive integers that is (also) expressed in a vertical diagram through the arrangement of the blocks on the worksheet. Efrem, too, suspects a regularity <442> through which we assume from the above analysis of the thematic development the fact that it results in the increase in the value of the yellow circles in a vertical direction. Both students however do not take their various reconstructed vertical systems as the central theme. The conversation takes another direction. Wayne persists in filling in line-by-line the third block as well, following the direction of the writing. He thereby further develops his number decomposition strategy (second warrant) in that he identifies the left-hand number of the two-digit number from the upper box and fills in the appropriate ten figure or allows it be filled in, in the left-hand box. Both are involved in producing the second conclusion: twelve <441, 443> says Wayne and Efrem fills in a ten <445> as well as a two <448>. Regarding the inference of the second conclusion it arises that Efrem enters the answers into his worksheet step by step and Wayne's appropriately inserted confirmations <446, 450> turn Efrem's statements into accepted inferences. As a presented layout the following argumentation is subject to the negotiation processes of the students from the point of view of the number decomposition framing (Figure 8.4).

Argumentation for Graphic-Framing

Within his graphic-framing Efrem's actions are accompanied by the production of two lines of argument. The first refers to the legitimacy of the number "1" to be filled in the right-hand box of the second row. The second line of argument concerns the numbers in the block on the third row. I address only the second argument here. Efrem points to the increase in the value of the numbers in the yellow circles in the following lines <343, 346, 348>. He does not count out the circles individually but merely hints at a counting process via his hand movements. He probably recognized the respective increases in the circles optically (diagram as warrant). The "visual presentability" of the circles at least makes the potential "countability" of the circles a possibility. This is seen as the backing for the inscriptional warrant. Consequently he draws a conclusion with regard to the number in the yellow circle (the inference of data 1 to conclusion 1, Figure 8.5). He brings the number of circles established in

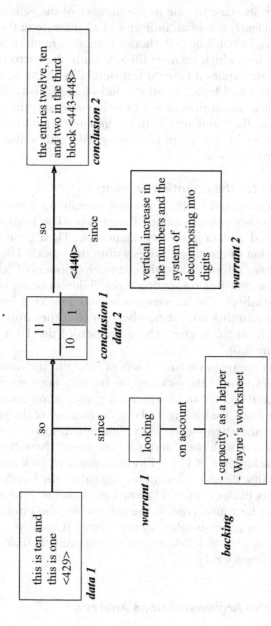

Figure 8.4. Argumentation structure of the number decomposition framing.

this way in the following second part of the argument, together with the right-hand box on the line into which a "1" should be written <438>. The step-by-step increase by one in the number of the yellow circles per row is to Efrem clearly less of an indicator of an increase in the ones to be entered in the right-hand box. Rather, it is more a numerical identification of the lines into which he must fill a one into the correct right-hand box. He therefore implies a general transferability of the answers in the second row right-hand boxes to all the following right-hand boxes. We can interpret this assumption as the non-thematic warrant of this third inference. Hereby the "similarities" in the geometric patterns of the rows could play a role within a graphic framing. Altogether the layout presented in Figure 8.5 arises.

Argumentation for the Counting-Framing

The jointly completed block of the row containing a twelve becomes the basis for another collective line of argument. The numbers filled in there are accepted as data in the argumentation. Their great similarity is in the presentation of the single boxes within the blocks. The presenting of particular boxes refers back to the persuasive powers of "closely looking" in the sense that this presenting a possibility of being able to work through the worksheet. The reference is seen as a backing for the following warrant. By counting on, the numbers for the other upper-boxes can be determined from the twelve. The repetition of the 10 is continually expressed (Figure 8.6).

Efrem speaks up again on the subject of filling in the numbers in the right-hand boxes (ones). He picks up on his own previous observations regarding this and creates an alternative argumentation about the rising sequence of the ones. He had figured out the number of the yellow circles in <434–438> and there—even if by different means—he had already made an association with the column for the ones. These thoughts function here as backing. In <454> Efrem appears to pick up on and to accept Wayne's first direct explanations. Regarding the boxes in the same line Efrem refers back directly to Wayne's explanations: the counting on as a pattern can be transferred to the column for the ones. During the conversation the warrant is taken as the theme. It conveys to Efrem the certainty of being able to follow Wayne's diagrammatic technique of filling in the gaps (Figure 8.7).

Summary of the Argumentations Analysis

I understand this episode as an interactive situation in which the beginnings of a framing difference are resolved during the evolving

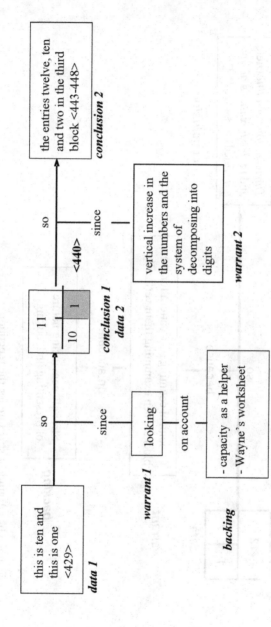

Figure 8.5. Argument structure for the emergence of "similarities."

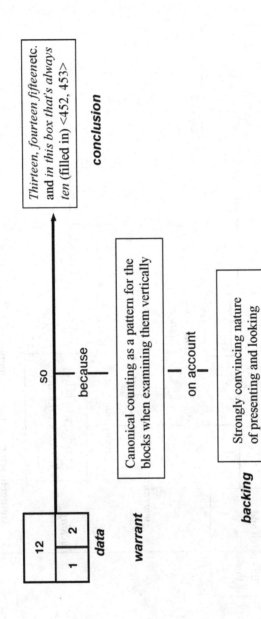

Figure 8.6. Argument structure for the repetition.

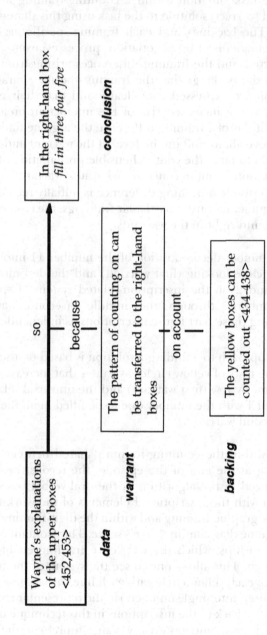

Figure 8.7. Argument structure for the filling in of gaps.

accounting practice. The result of the negotiation of meaning is the interpretation of the task situation within a counting-framing and the conjointly produced (correct) solution to the task using this shared manner of
interpretation. The backings and each framing-specific argumentation
remain non-thematic in the negotiation processes using a reflexive
accounting practice and the framing differences arising from them. It is
precisely within the backings that the framing-specific characteristics of
an interpretation are expressed very clearly and that their explicit thematic nature holds an increased risk of the interaction breaking down.
Under the conditions of a framing difference the argumentative clarifica-
tions appear, above all, to shift on the level of the warrant and the focus of
the debate points towards the comprehensible presentation of the "core"
of the argumentation "Data so conclusion, because warrant."

In the above episode a framing difference is initially reconstructed. In
the relatable argumentations of particular framings, the connection to the
inscriptions is found right in the warrants:

- Wayne's number decomposition of the number 11 into 10 and 1 is
  based on close looking (first warrant) and the decomposing of 12
  into 10 and 2 on the inscription-related system of splitting two-
  digit numbers "through the middle" (second warrant) and
  interpreting these part figure inscriptions as independent numbers
  again.
- Efrem's approach to a finding a solution is based on the interpretation of the rows of yellow circles as values that increase from top to
  bottom row by row (first warrant) and the universal relation of this
  increase of 1 with the number one to be filled in all the right-hand
  boxes (second warrant).

In contrast, within the counting-framing shared between Efrem and
Wayne occurring at the end of the episode, the reconstructed warrants
become thematized in the algorithm of the oral-verbal based counting.
The connection with the inscriptional elements of the worksheet diminishes. Within the graphic-framing and within the digit-framing the parts of
inscriptions become diagrams in Peirce's sense. They contain object-repre-
sentamen relationships, which the respective framing enables interpretants to be created. This allows one to see the system of the inscriptional
object (*increase* of each yellow circle with each line, *decomposing* of two-digit
number inscriptions into single numbers) in the representamen. As it is relatively easy to refer back to the inscriptions in this technique using appropriate verbal indications and gestures, we can comprehend the consistence
and harmony of the argument "data so conclusion, because warrant" that
is created within a specific framing. This comprehension is made easier by

the fact that within many mathematical framings it is possible to identify the sign with its object.

There emerges a "diagrammatic argumentation" which reveals a reflexive-rational handling of framing differences. These disparities do not always necessarily lead in the interpretation of the situation to interactive conflicts. The inclusion of inscriptions in mathematical subject development enables diagrammatic argumentation through which we "see" the "cause" of a mathematical connection as it were. In this sense we can understand Bruner's epigram "We *see* cause" at the beginning of this chapter.

THEORETICAL CONSOLIDATION: NARRATIVE ACCOUNTING PRACTICES

In this section I will take a more in-depth look at why such diagrammatically based argumentation within negotiation processes possess such great powers of persuasion. Primary school children in interactive problem solving situations produce (collective) argumentations, which are narratively structured. These narrative theses refer to both the performance and the accountability of an action. Thus, not only is the method of problem solving narratively structured but also the argumentation contained in it. For a detailed explanation we would have to bring in narrative theoretical terms such as, for example, the "plot" and the argumentation theoretical considerations for example topos or locus. In the following, I discuss above all, the effectiveness of diagrammatically based argumentation in such narrative accountings practices.

There are four characteristics of narrative production: sequentiality, indifference between the real and the imaginary, unique way of managing departures from the canonical, and dramatism (Bruner, 1990). In narration the appropriate specific order of events is important. One event follows another and cannot, in the sequencing of the presentation, simply be changed. Furthermore successful stories stand out in that one cannot differentiate whether they are true or made up. Moreover it is typical to endeavor refering unknown facts back to known facts and to present unusual occurrences as normal. The dramaturgy of presentation ensures that with a good story one does not want to stop listening.

In the context of narrative argumentation, the four characteristics of narrative production from Bruner's canon, in particular the first and third, are of interest. The first point includes the formal structural aspect of an argumentation: the order of the actions is given at the same time as it is dealt with "logically." That which happened first is told first (sequencing). The third point describes the motive of why we need to argue. The task, which the children must work on, differs up to this point. It would be solvable if one could refer back to and already "had" a problem solving method. "The function of the story is to find an intentional state that mitigates or at least makes comprehensible a derivation from a canonical cultural pattern. It is this achievement that gives a story verisimilitude. It also may give it a peace-keeping function" (Bruner, 1990, p. 49). As in narration, which distinguishes itself through a specific sequence of actions, in the mathematics lesson, the newly interpreted problem-solving situation under the activation of an accustomed framing would be transferred into a "canonical" definition of a situation appropriate to this framing.

Even if one wanted to ascribe to Efrem the fact that he is still looking for a profitable interpretation of the task, one can ascribe to both students that they explicate the apparently canonical way of seeing the nature of the task particularly through the awarding of inscriptional warrants. Through a reflexive accounting practice, the foundation for this claim of a canonical perception is simultaneously contained in the realization of the problem solving process. It does not even cross Efrem's and Wayne's mind that others working on the task might interpret it differently and therefore need a different justification for their suggested answers. Certainly supported by the persuasive power of their diagrammatically based argumentation, they assume an existing "canonical" understanding of the task.

The persuasive power of the kind of narrative accounting practice lies for both students in the diagrammatic use of inscriptions, which can be manipulated according to a set of rules. These rules manifest themselves in the sequence of actions, which from the viewpoint of accountability of the actions in their unalterable order represent the "logic" of the action. The legitimacy of an argumentative inference comes in the warrant to the language. Colloquially, the "reasons" for the argument are articulated within it. In the mixing of verbal-oral actions and visual-manual actions, typical for mathematical interaction, two dimensions are made visible: (a) the presented episode in the production of diagrams on a visual-manual level and (b) the legitimacy of the sequence of actions as it were ("We see cause").

The potential narrative accounting practice lies in the imparting or moderation of framing differences and thereby in the possibility of being able to convince someone of one's own perspective within a different framing. A tool for the production of such practices is the diagram that allows a diagrammatically based argumentation to be built—in the ascribed (determined through writing) manifestation of a specific sequence of actions. The argumentation demonstrates above all on the level of the warrant this diagrammatic aspect. The visual-manual character of this explanation seems to radiate particularly great powers of persuasion (verisimilitude). This lies in the fact that interpretants can be produced within many framings of mathematical inscriptions, which allow the mathematical object to be identified with its sign.

The exclusively language-based interpretation of an abstract mathematical object is often superseded by the inscriptional manipulation of its sign. This is a change of direction to "diagrammatical materiality" (Dörfler, 2006, p. 215), which would completely correspond to the "main business of the mathematicians," the option of a facilitation for mathematics students. As the end of the episode shows, such a shift in the thematic development of the conceptual idea of a decimal place value system to a diagrammatical manipulation of a graphic illustration, does not however necessarily lead to a continuing understanding among the students: within the counting framing finally accepted by Efrem and Wayne, the idea of a decimal number presentation largely disappears.

METHODOLOGICAL COMMENTS: ON THE FUNCTION OF THE EXAMPLE

The example takes on various functions in the current text. The concrete episode serves as an illustration. In selected parts of the interpretation, the terms introduced are clarified. Here we could have chosen a fictitious example and thereby could have possibly explained some of the aspects of the concept formation more concisely. Through use of part of a "real" lesson, in view of the illustration of terms, an additional analytical aspect of the essay is brought in. At least, by using the example we are shown that a theoretical approach is introduced here, which also works with terms that are empirically rich in substance. It is shown how, in a section of a lesson that actually took place, the developing terms that are empirically rich in substance, can be used and thereby particular aspects of this real lesson are describable and nameable.

How far readers can see the possibility of being able to usefully apply these terms to other lessons in a single episode will depend on their judgment and experience. In view of the research carried out, I see this as a convincing possibility. In several empirical projects, the theoretical elements explained here have been developed. Comparative analyses have been methodologically carried out, whereby in these projects the "Constant Comparative Method of Qualitative Analysis" (Strauss & Corbin, 1990) has been taken up and been fitted to the particular object of investigation. According to this methodical approach, one can assume that a theory and the terms (empirically rich in substance) contained within it, could claim a global validity, if empirical theoretical development of interpretations of diverse and potentially contrasting episodes is carried out: "there will be a wider applicability of the theory, because more and different sets of conditions affecting phenomena are uncovered" (p. 190).

The function of the example changes when the reconstructed arguments are compared to each other. One may recognize how here the above named "Constant Comparative Method of Qualitative Analysis" has an effect. The identification of similarities and differences in the reconstruction of the three arguments "leads" us almost to make theoretical assumptions about how one can explain these similarities and differences. Although it is a matter here of a comparison of argumentation from one single episode, we can claim that the way in which the students bring in, through the inscription of the circles-illustration and the inscription of the decimal number diagram, their implicit argumentation, allows us to recognize a substantial feature of reflexive accounting practice in mathematics lessons in the primary school. For the continuation of interaction within an emerging framing difference it is increasingly referred back to a diagrammatical use of (mathematical) inscriptions, in which an identification of the mathematical object with its sign is made possible within many framings. The rule-governed manipulation of the sign enables conclusions to be made about the relevant objects and insights—depending on each individual framing. The ascribed framing difference does not have to be recognized by the participants. It may even be that under these conditions this difference in the thematic development can be disregarded.

APPENDIX: TRANSCRIPTION CONVENTIONS

The transcript is divided into three columns.

1. In the first column the lines are numbered in sequence. The line numbers here refer to their original sequence. As the breaks in the lines differ, only the first line number is given for each utterance. In the text, these numbers are given in pointed brackets.

2. In the second column the speaker is referred to by name; the teacher is shortened to L. If there is an overlapping of speech, the transcript continues in script format. *Simultaneity* is noted by a pointed bracket (<) in front of the name.

3. In the third column verbal utterances in as far as they can be understood are given in normal font. The verbal utterances are augmented by action, expression and gesture which are briefly noted.

The following symbols and formatting are given further differentiation.

/ \ -	voice is raised lowered or remains the same
. .. ...	Pauses in speech; the number of dots denotes the length of the pause (per dot approx. 1 second) Transcript during the passages of speech contains no punctuation
S p a c e d o u t	said slowly
Bold	words said with emphasis
+	end of a given utterance, gesture, facial expression, action, breath (word) bracketed words are not completely intelligible (inaudible) completely unintelligible utterance

NOTE

1. In this chapter I frequently refer to research results which were achieved in with close co-operation of my colleagues. The respective reference to the further reading section will clarify who is meant by "we" in particular cases.

REFERENCES

Bruner, J. (1986). *Actual minds, possible worlds.* Cambridge, MA: Harvard University Press.

Bruner, J. (1990). *Acts of meaning.* Cambridge, MA: Harvard University Press.

Bruner, J. (1996). *The culture of education.* Cambridge, MA: Harvard University Press.

Dörfler, W. (2006). Diagramme und Mathematikunterricht. *Journal für Mathematikdidaktik, 27,* 200–219.

Fetzer, M. (2003). *Interaction in collective wriring processes in early mathematical learning.* Bellaria, Italy: ERME.

Garfinkel, H. (1967). *Studies in ethnomethodology.* Englewood Cliffs: Prentice-Hall.

Goffman, E. (1974). *Frame analysis. An essay on the organisation of experience.* Cambridge, MA: Harvard University Press.

Habermas, J. (1985). *Theorie des kommunikativen Handelns.* Frankfurt a/M: Suhrkamp.

Peirce, C. S. (1978). *Collected papers of Charles Sanders Peirce.* Cambridge, MA: Harvard University Press.

Pimm, D. (1987). *Speaking mathematically: Communication in mathematics classrooms.* London: Routledge.

Roth, W.-M., & Mc Ginn, M. (1998). Inscriptions: Toward a theory of representing as social practice. *Review of Educational Research, 68,* 35–59.

Strauss, A., & Corbin, J. (1990). *Basics of qualitative research. Grounded theory procedures and techniques.* Newbury Park, CA: Sage.

Toulmin, S. E. (1969). *The uses of argument.* Cambridge, England: Cambridge University Press.

EDITOR'S SECTION COMMENTARY

Wolff-Michael Roth

In the three preceding studies, we see students of different ages engage in mathematical endeavor not merely for the purpose of filling a sheet of paper to be submitted to the teacher, but for the express purposes of articulating their individual|collective thinking to others and in ways already presupposed to be intelligible to an audience. Although in each case, this audience includes the teacher, it consists primarily of peers who are at approximately the same level in their mathematical development and at the same competency level with respect to talking mathematics.

From a traditional perspective on learning, there would little difference between what students do, say, and write when they address the teacher only—homework, test, examination—and when they address a wider audience. This is so because all a student is thought to do is empty out the contents of their minds, using language as a medium to carry the information from within themselves to without.

A very different perspective on this issue comes both from cultural-historical activity theory and from literary scholarship. Thus, in cultural-historical activity theory, the object/motive consists of the materials, on the one hand, and the ultimate product and its use in the collective, on

Mathematical Representation at the Interface of Body and Culture, pp. 245–248

the other hand (Roth & Lee, 2007). As all other irreducible moments of the activity system, the anticipated outcome and its future use mediate the productive process and therefore leave traces in the produced entity. Similarly, from a literary perspective, any utterance—which can be a word, sentence, poem, or a novel—is made in view of the audience and in view of its anticipated response (Bakhtine/Volochinov, 1977). Again, the anticipated recipients and their responses mediate what and how an individual or collective human subject communicates.

Both these perspectives therefore suggest that we need to understand what we see in the three preceding chapters as products that not only characterize the speaker(s) but also the addressee, which includes teacher and peers. There is therefore a clear orientation toward the other, who can only understand the speaker when the latter already uses a common language and common grammar. That is, far from building intersubjectivity in such instances, students always and already have to arrive with a high degree of intersubjectivity so that they can participate in the classroom conversations, be they at the elementary, middle school, or university level. More so, the speakers and audiences have to orient toward some common rules, which they, in their actions, both reproduce and transform. It is only when we think of these situations in terms of dialectical production—that is, reproduction and transformation—that we come to grips with the kinds of transformations observable in mathematics lessons oriented toward establishing common norms and practices (Roth & Lee, 2006).

In all three chapters, the students focus on and talk about inscriptions, graphs, arithmetic sequences, or differential equations and their solutions (graphs). Inscriptions are material and therefore subject to perception and the visual aspects of cognition. From a dialectical materialist perspective, we must not reduce these perceptual aspects to the verbal, for we would otherwise loose essential aspects of thinking (McNeill, 2002). It is precisely when we begin to think the verbal and perceptual aspects as mutually constitutive, irreducible expressions of the same higher order communicative unit (which they can only express in a one-sided way) that we can come to grips with understanding communication over and about inscriptions. Moreover, perception and imagery draw on the body and its ways of orienting in the world, as shown in chapter 5. Together with the gestures that also express the perceptual moment of thinking and communicating, we see—even if the authors do not explicitly write about it— that there is an essential embodiment aspect to communicating in collectives and in the establishment of norms of communication.

It was Lev Vygotsky (e.g., 1986), the father of cultural-historical activity theory, who alerted us to an integrating unit of analysis. This unit, which he presents only as a sketch, sublates what are frequently seen to be dichotomous phenomena. Thus, in his work on communication, he clearly articulates the dialectical (dialogical?) relation between individual thinking and its expression in the communication with others in the social sphere. Thus, he clearly sublates the dichotomy of the individual and the collective. Moreover, he sublates the dichotomy between the body and the mind, even if sketchily. McNeill (2002) shows, for example, how Vygotsky's framework has to be expanded to include speech and gestures as two irreducible moments of communication. Moreover, Vygotsky considers emotions, largely considered to be bodily phenomena, and thought as expressions of one phenomenon rather than as two phenomena that affect each other from the outside. Vygotsky considered the segregation of thought and emotion as useless, as it "makes the thought process appear as an autonomous flow of 'thoughts thinking themselves,' segregated from the fullness of life, from the personal needs and interests, the inclinations and impulses, of the thinker" (p. 10). It is only when thought and emotions entertain an *inner* connection that we can come to understand how they mutually affect one another and volition.

Finally, emotions are not individual phenomena, but they are irreducibly social as well (Collins, 2004). We learn to relate bodily states to emotion talk in communication with others, and thereby become emotionally social beings in and through our participation in societal activity with others. Furthermore, in collective endeavors, individual and collective emotion stand in a mutually constitutive relation, as collective emotion cannot exist but in and through the individual bodies and yet, collective emotion does mediate the emotions of individuals. The role of emotion in holding together communities of practice has not yet been explored in the literature on mathematical learning and cognition (but see Roth, 2007). There is therefore a widely open area of future research on the collective production of mathematical practices and norms.

REFERENCES

Bakhtine, M./Volochinov, V. N. (1977). *Le marxisme et la philosophie du language: essai d'application de la méthode sociologique en linguistique* [Marxism and the philosophy of language]. Paris: Les Éditions de Minuit.

Collins, R. (2004). *Interaction ritual chains*. Princeton, NJ: Princeton University Press.

McNeill, D. (2002). Gesture and language dialectic. *Acta Linguistica Hafniensia, 34,* 7–37.

Roth, W.-M. (2007). Mathematical modeling "in the wild": A case of hot cognition. In R. Lesh, E. Hamilton, & J. J. Kaput (Eds.), *Foundations for the future of mathematics education* (pp. 77–97). Mahwah, NJ: Lawrence Erlbaum Associates.

Roth, W.-M., & Lee, Y. J. (2006). Contradictions in theorizing and implementing "communities." *Educational Research Review, 1,* 27–40.

Roth, W.-M., & Lee, Y. J. (2007). "Vygotsky's neglected legacy": Cultural-historical activity theory. *Review of Educational Research, 77,* 186–232.

Vygotsky, L. S. (1986). *Thought and language.* Cambridge, MA: MIT Press.

PART C

STEPS TOWARD RETHINKING
MATHEMATICS EDUCATION

EDITOR'S SECTION INTRODUCTION

Wolff-Michael Roth

For me personally, Grammar school (Ger., *Gymnasium*) mathematics has been a drab experience. For the life of me, I often barely made it through the curriculum, although, without that I could really put my finger on it, there were things in my life that I know today would have been resources for successful mathematics experiences at a much earlier stage than when they eventually came. Thus, for example, I found it difficult to do, in the textbook-oriented lessons of my youth, a "problem" such as $a^2 - 5a - 14$. Yet I also clearly remember many a sleepless night that I passed figuring out mathematical things in my head while trying to get myself sleepy. I remember to this day the experience I had at about 14 years of age trying to figure out how to calculate 29 x 29 in my head and without access to pencil and paper (calculators did not yet exist in those days). Lying on my back in the dark, I was visualizing how the calculation would unfold on paper. But then—because it became difficult to keep in mind the different numbers to be added up and their horizontal shift—I tried to figure out how I could get from knowing that 30 x 30 = 900 to the solution of the problem I had posed. I envisioned that if I took 29 x 30, I would have to take 1 x 30 away from the 900. Then I realized that I could take 1 x 29 away and thereby reduce the second multiplier. So I knew my result would be 900 − 30 − 29, which I could figure out, without problem, to be 841. But then it dawned on me that this is the same as 900 − 30 − 30 + 1. I asked myself if this had to do with the two 30s that I had started out with.

This led me to postulate that $29^2 = (30 - 1)^2 = 30^2 - 2 \times 30 + 1$. I thought it would be neat to try this out with other numbers, such as 19^2 or 39^2. And then I did the same with numbers that were closer to the lower tenth (e.g., 24^2). I wasn't algebra savvy at the time so my generalization was imagistic or verbal: take the next higher tenths and square it, subtract (add), then subtract (add) twice the tenths, and add 1.

Today I know that my teachers failed to recognize the potential in such mathematical engagement and failed to utilize it not only for my own benefit but for that of the class as a whole. Moreover, I was interested in mathematical problems, but school mathematics did everything to kill this interest. Today I realize that I was so persistent because it was "my" problem, it was a real problem that I pursued, and it was something that today I would recognize as a mathematical intention or object/motive. I had bought in to this motive, and there resulted a high level of satisfaction when I had figured out the pattern. It was tremendous and self-motivating—but I never felt like this in the mathematics class. I imagine that there are many students who experience mathematics as a drab but who might become excited when they come to figure out patterns to truly problematic issues, that is, for example, to self-chosen questions. I am sure that there are ways in which we can think about and organize school mathematics to make a difference. Given the issues raised in this book, there must be ways in which we can rethink and reconfigure school mathematics that allow for the integration of embodiment and culture in the way the analyses in the previous two parts present them.

In this Part C of the book, I bring together three chapters all of which propose novel perspectives on thinking and doing mathematics education or training mathematics teachers. Though I have not thought sufficiently about this, my research in a variety of workplaces suggests that a lot of mathematical learning may go on precisely when it is *not* the object/motive of the task. There is also experimental evidence about the learning of formal subject matter by the founder of cultural-historical activity theory, Alexei N. Leont'ev. In the German version of his *Activity, Consciousness, and Personality* (Leontjew, 1982) there is a chapter that does not appear in the English translation. In it, Leont'ev tells the story of an experiment that he and his collaborators conducted in the science museum of Charkov, where there was an opportunity for young people to learn about airplanes while building models thereof. The researchers found that students were not at all interested in learning about the physics of flying and focused on building nice models. As soon as the object/motive of the task was changed to building planes that passed a certain line in the distance, the young pioneers significantly increased their questioning of instructors and consultations of formal physics concepts after their first prototypes failed. More so, the average daily attendance increased from about 6 individuals to over 40 individuals after the change. That is, more students studied formal physics concepts for periods of times an order of magnitude larger than before. I would hope that we could find forms of engagement for students today that similarly engage them and their interest in mathematical inquiry.

Unfortunately, I just have to look around myself at work and see how little there is done to change the training of mathematics teachers. A lot of the classes turn out to be responses to "How do I get my students successful on standardized examinations?" and some lip service to constructivism. But mathematics teacher education does not have to be so boring. All three chapters in this section of the book insist on this point.

In chapter 9, Brent Davis is concerned with mathematical accounts and accountability. In contrast to the other chapters assembled in this book, his chapter "So What...?" precisely focuses on teachers' explorations of mathematics through metaphors. He argues that given the relative novelty of research into the images, metaphors, analogies, gestures and exemplars that are manifest—implicitly and explicitly—in mathematics, it is not surprising that discussions of implications for classroom action are limited and underdeveloped. As well, when and where applications of emergent insights are evident, they are often perceived to be in conflict with entrenched structures of curriculum and expectations of schooling. Against this backdrop, Davis represents teachers' perspectives on the issues presented in this book generally, specifically as they engage with the question of what this research might mean for their practice. Speaking as an educational researcher and teacher educator who has worked with practitioners for nearly a decade around this question, Davis also addresses the matter of how the issues and topics might be framed for prospective and practicing teachers.

My own experiences of mathematics were textbook driven, doing the tasks that followed an exposition of some technique, such as how to factor the polynomial $a^2 - 5a - 14$. There never was a question about any other way of thinking about the pertinent issues then getting the preformatted answer. Although many moons have gone by since I attended Grammar school, I have seen—from close-up during my years as a classroom teacher (science, mathematics, computer science)—that little has changed in the way mathematics was taught. There was often a conflict between my physics students, who learned calculus, using graphics calculators, months prior to the treatment of these topics in their mathematics course. Here, we dealt with multiple representations of physical phenomena students were investigating. How do multiple representations come together and participate in mathematics learning and insight? Very different from the frequently formula- and fact-driven approaches in mathematics classrooms, Ian Whitacre, Charles Hohensee, and Ricardo Nemirovsky look for the growth of mathematical understanding while students engage with physical phenomena. In chapter 10, "Expressiveness and Mathematical Learning," they provide answers to their framing question by observing the engagement of students in an environment where multiple representations are available. The primary phenomenon that the authors examine is how aspects of these multiple representations come to be incorporated into streams of bodily activity, exhibiting features of expression. As a result of their analyses, they propose the metaphor of the student-as-artist to capture the ways in which students bring together and fuse different representational elements in the singular lifeworld of their experience. Their study in fact sets readers up for the subsequent chapter.

A lot of mathematics teaching was, and continues to be, directed toward doing textbook-based tasks that often lead to little mathematical *understanding* even though students may be successful on examinations. The studies of mathematics in everyday life showed that there is a low to inexistent correlation between school mathematics (number of courses, achievement) and how well people solve problems in their lives, especially in situations that really matter to them, that is, situations that are truly problematic (e.g., Lave, 1988). Some questions are never asked in schools, including "What is mathematics?" and "What is its nature?" Rafael Núñez suggests in chapter 11 that these questions have been mainly addressed in the philosophy of mathematics, logic, and in mathematics itself (e.g., meta-mathematics). Research in the teaching and learning of mathematics and their cultural practices, as well as other behavioral investigations of human mathematical performance, have the potential to provide rich insight into the question of the nature of mathematics. Most of these studies, however, have addressed questions of teaching, learning, or practice of mathematics, where "mathematics" usually is a pre-given content and therefore have left the very body of knowledge of mathematics intact (e.g. theorems, axioms, definitions, notations). Núñez shows that this does not have to be the case. The empirical study of human imagination, abstraction, and conceptual systems can provide rich insight into the question of what is mathematics and what makes it possible. He illustrates his arguments by looking at contemporary theories of conceptual mappings (e.g., metaphors, blends) and the empirical investigation of human gesture-speech co-production, when they are applied to specific domains of mathematics, such as continuity in calculus.

REFERENCES

Lave, J. (1988). *Cognition in practice: Mind, mathematics and culture in everyday life.* Cambridge, England: Cambridge University Press.

Leontjew, A. N. (1982). *Tätigkeit, Bewusstsein, Persönlichkeit* [Activity, consciousness, personality]. Köln: Studien zur Kritischen Psychologie.

CHAPTER 9

AND SO...?

Brent Davis

Given the relative novelty of research presented in the previous chapters, it is not surprising that discussions of implications for classroom action are limited and underdeveloped. As well, when and where applications of emergent insights are evident, they are often perceived to be in conflict with entrenched structures of curriculum and expectations of schooling. Against this backdrop, I represent some teachers' insights around the sorts of issues raised in this book. Speaking as an educational researcher and teacher educator who has worked with practitioners for a decade around these matters, I also discuss how issues and topics might be framed for prospective and practicing teachers.

INTRODUCTION

For the past ten years, I have been working with different groups of practicing teachers to understand how we, as mathematics educators, might make sense of and apply the sorts of ideas presented in this book. Specifically, following recent research into the embodied bases, figurative aspects, semantic qualities, and distributed nature of mathematical concepts, we have been attempting to represent, interrogate, and elaborate

Mathematical Representation at the Interface of Body and Culture, pp. 257–273
Copyright © 2009 by Information Age Publishing
All rights of reproduction in any form reserved.

their mathematics by studying what teachers know, how they know it, and how they enact it in educational settings.

In terms of the grander context, this work ties in with what is perhaps the most prominent topic in contemporary mathematics education research: teachers' disciplinary knowledge. For decades, investigations of this issue were organized around unproductive efforts to find correlations between courses teachers take and scores their students earn. More recently, attention shifted toward what Ball, Hill, and Bass (2005) call "job analysis." This approach involves close examinations of the skill requirements of teaching as contrasted with skills needed, for example, in nursing, engineering, or architecture. As Ball and her colleagues note, teachers must be adept at unpacking concepts, organizing examples, and interpreting student actions. These competencies require specific disciplinary knowledge about, for example, how mathematical topics are connected, how ideas follow from and anticipate others, and what constitutes a valid argument.

An assumption beneath much of this research is that there is, in fact, a body of knowledge that can be isolated, specified, organized, taught, and (ultimately) measured. And, indeed, considerable progress has been made over the past several years in identifying critical and distinct elements of teacher's disciplinary knowledge of mathematics. Unfortunately, as more studies are conducted, the phenomenon is proving to be as elusive as ever. It is revealing itself to be vaster, more distributed, and more tacit than most had imagined—reflecting many of the discussions offered in the preceding chapters. As well, it is proving to be a moving target; as I develop in this chapter, it changes when it is studied.

It is the realization of these confounding aspects—the depth, the breadth, along with the manner in which teachers' knowledge is continuously and spontaneously elaborated—that orients the writing. This chapter is, in effect, a report on what can happen when teachers are invited to study such issues as the metaphoric basis of mathematical knowledge and the distributed nature of mathematics knowing in the context of learning and teaching the discipline. As I develop, it is an ongoing exploration that prompts a recasting of teachers' mathematics more in terms of an *attitude toward* knowledge than a *body of* facts, and teachers more as collaborators in than consumers of educational research.

KNOWING WHAT WE KNOW

This move to highlight attitude or disposition as a vital part of teachers' disciplinary knowledge is supported by Chevallard's (1991) observation that, with few exceptions, bodies of knowledge are developed to be used, not to be taught. With a focus on utility rather than meaning, experts across domains often give little attention to the associations and supports that were employed by others to generate important insights and processes. This point is not surprising. Not only is conscious awareness of such elements typically unnecessary for experts, but also it can actually be debilitating.

This certainly seems to be the case for disciplinary mathematics, which is precisely why a teacher of mathematics must adopt a very different attitude from a researcher of mathematics—or so I will argue here. As it turns out, it seems that most teachers have a wealth of enacted mathematical knowledge that they do not necessarily know that they know, and this point has only recently been brought into sharp relief with work into the embodied and figurative bases, along with the situated and distributed nature of mathematics knowledge.

A cogent demonstration of this matter occurred several years ago when, in a project involving a cohort of 25 K–12 teachers and my colleague Elaine Simmt, the question "What's multiplication?" came up. The immediate response was that multiplication was "a grouping process" or "repeated addition," and after a little probing it seemed that these descriptors more or less summed it up for most present.

Well aware that many other interpretations of multiplication are invoked in grade school curricula, Elaine and I pressed the matter by noting that grouping and repeated addition make sense when working with discrete objects and whole numbers, but these "definitions" become more and more troublesome as other number systems and applications are encountered. We asked participants to consider how they frame multiplication of fractions, integers, irrationals, polynomials, matrices, vectors, and so on, in the process hoping to gain access to some of what might be enacted, but perhaps not made explicit in their classrooms.

Over the next hour or so, the teachers chatted and debated, eventually generating the following list: repeated addition; grouping; sequential folding; many-layering (literal meaning of "multiply"); opposite/inverse of division; intermediary of adding and exponentiating; the basis of proportional reasoning; grid-generating; dimension-changing; scaling; stretching or compressing of number line. Notably, this result has proven to be remarkably robust; under similar circumstances, several other groups of pre-service and practicing teachers have since produced very similar lists, both in terms of entries and order. Indeed, when I was working a few years ago with a different group of a dozen middle school teachers, virtually the same set of interpretations (with the addition of branching, as well as number line rotation for the multiplication of signed numbers) was generated. In a nutshell, while incomplete in terms of the range of definitions, algorithms, images, and metaphors of multiplication that arise in the K–12 experience, I believe the list to be strongly suggestive of the principal interpretations.

The more recent event, involving the middle school teachers, was interesting for another reason besides the similarity of the result. I was also struck by the multi-layered and very emotional responses of the teachers—one aspect of which was collective surprise at both the range of interpretations and the fact that, although never previously encountered in this form, participants were all "aware" of them and, in fact, "used" them in their teaching. Alongside the surprise was a palpable pleasure at the process of becoming more attuned to what they already knew—a detail that takes me back to the points at the start of this section that teachers' mathematics is more about an attitude toward knowledge than a body of facts, given that it may be more a matter of tacit associations than explicit information.

At the same time, however, there was a seemingly discordant emotional response that came across as a sort of collective apology for the list of interpretations, expressed by one of the eighth-grade teachers who explained, "We just use these things to get at the idea of multiplying. It's not really the math." Even more disconcerting was her follow-up question, met with nods of approval around the room: "So if multiplication isn't repeated addition, what's the right definition?"

This was one of those moments when what had been taken-as-shared is revealed as anything-but-shared. It was experienced as a rupture, as the pleasure of grappling with the complexities of a concept ran up against the need to simplify that concept for teaching—or more cogently, the desire to frame teachers' mathematics as an attitude toward knowledge crashed into cultural tendency to view mathematics as a collection of facts. As we looked at one another we realized that we were thinking in very different ways about some fundamental issues, including the nature of mathematics knowledge and the nature of mathematics knowers.

COMPLEX KNOWLEDGE

One of the aspects that I most appreciate about the teachers with whom I work is their willingness to suspend disbelief. Of course, when the group just mentioned first began meeting together, the usual worries around test scores, curriculum coverage, and math anxiety were front and center. However, on the assurance that the research was about improving practice by focusing on what teachers *do* know rather than what they *don't* know, they seemed willing to set those ever-pressing concerns aside—at least during the research sessions. In brief, there was a clear (and stated) acknowledgment that they were willing to devote energies to matters that might not have immediate and obvious practical import.

It was thus that the "what's the right definition?" moment was something of a crisis point. On the one hand, it served as a powerful confirmation that we were indeed finding ways to render tacit mathematical knowledge more explicit. On the other hand, it seemed to culminate with a realization that we were wandering too far afield of the pragmatics of teaching by highlighting that none of us seemed to have a good grasp on something as basic as multiplication. Not wanting to position myself as the authority, I was happy to follow along with a participant's suggestion to consult a few online dictionaries. We started with the glossary at math-propress.com, where multiplication was defined as the "basic arithmetical operation of repeated addition." Unfortunately, we had just demonstrated that meaning could be woefully inadequate.

The dictionary at mathresouces.com, in contrast, began its lengthy definition of multiplication as "a binary arithmetical operation defined initially for positive integers in terms of repeated addition, by which the product of two quantities is calculated, usually written $a \times b$, $a \cdot b$, or ab." When this sentence was read aloud, the word "initially" caught the attention of at least a few participants, with one commenting, "So this dictionary is suggesting a more open definition of multiplication." The phrase *open definition* was the trigger for some animated discussion for how to think about the learning and teaching of mathematical concepts. Rather than seeking to finalize definitions, the notion helped us to think about how a concept such as multiplication can be introduced in a manner that allows for it to be elaborated in ways that preserve but extend established insights. That realization, in turn, prompted the group to be somewhat more systematic in its examination of the images, metaphors, and methods that pop up in K–12 curricula, seeking to map out when they arose and how they might be connected and differentiated.

One of the results of this aspect of the work is presented in Figure 9.1,[1] in which a "landscape" of the concept of multiplication as it arises and is elaborated is presented. This image has actually evolved considerably over the year since we began to organize our thinking. I do not go into much detail around its actual development other than to highlight the categories listed across the top. These were invented to cluster and distinguish the varied interpretations according to core images and metaphors. In effect, these categories point to some of the more deeply inscribed interpretations of multiplication, which are rooted in repeated assembling, steady movement, steady growth, partitioning, and composing. Mapped against some of the applications of multiplication that appear in K–12 curricula (in the box on the left), a sense of coherent elaboration is

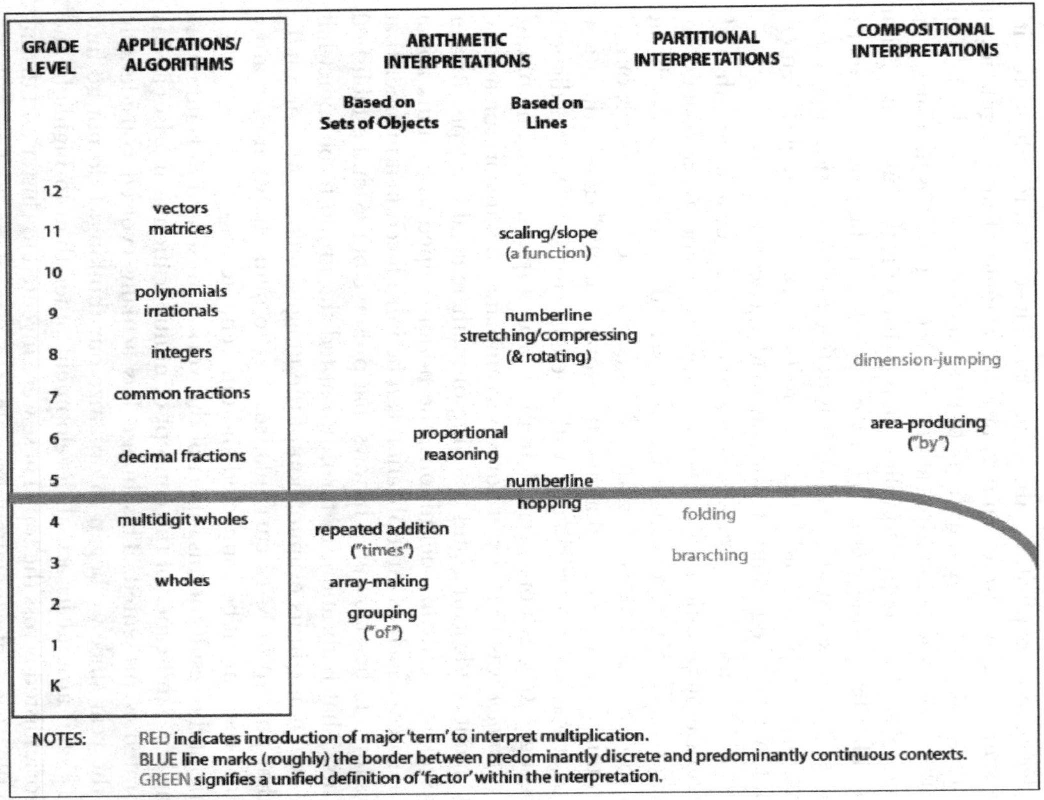

GRADE LEVEL	APPLICATIONS/ ALGORITHMS	ARITHMETIC INTERPRETATIONS		PARTITIONAL INTERPRETATIONS	COMPOSITIONAL INTERPRETATIONS
		Based on Sets of Objects	Based on Lines		
12	vectors				
11	matrices		scaling/slope (a function)		
10	polynomials irrationals				dimension-jumping
9	integers		numberline stretching/compressing (& rotating)		
8	common fractions				area-producing ("by")
7		proportional reasoning			
6	decimal fractions		numberline		
5			hopping	folding	
4	multidigit wholes	repeated addition ("times")		branching	
3	wholes	array-making			
2		grouping ("of")			
1					
K					

NOTES: RED indicates introduction of major 'term' to interpret multiplication.
BLUE line marks (roughly) the border between predominantly discrete and predominantly continuous contexts.
GREEN signifies a unified definition of 'factor' within the interpretation.

Figure 9.1. A landscape of MULTIPLICATION.

presented—even if the elements of this elaboration are rarely, if ever, made explicit when teaching.

The notions of "open definition" and "landscape," which were proposed by participants themselves, have been very useful in surfacing some tensions in individual and collective understandings about mathematics within the group. For example, as the uncomfortable moment around "the right definition" of multiplication illustrates, in teaching there is something of a lived tension between mathematics as a fixed, explicit, and clearly defined domain and mathematics as an evolving, largely tacit, and openly defined phenomenon. And, as the landscape in Figure 9.1 hints, the deeply entrenched assumption that mathematics has a pristine and hierarchical structure (hence the linear, incremental curriculum) is manifest alongside a messier, networked structure in the actual contents and methods of teaching.

It would be misleading to suggest that this particular issue (i.e., hierarchical vs. networked structure of mathematics) came up naturally in the course of conversation with the teachers. It was a topic that I introduced quite deliberately shortly after the group began to map out their landscape of multiplication. Prompted by contemporary work on complex systems and their decentralized network structures, I felt it would be productive if I inserted some information on network theory and mathematics education (e.g., Mowat, 2008). Specifically, I hoped this emerging work might provide a means to foreground popular assumptions about the structure of mathematics (i.e., as pristine and hierarchical) while offering alternatives (e.g., scale-free, decentralized network). To illustrate the significance of this reframing for teaching, a hierarchy would suggest a sequential and logical curriculum that moves from simple foundations to complicated constructions, whereas a scale-free network would point more toward dwelling with highly connected ideas (such as multiplication) and exploring local networks, with the conviction that a deep knowledge of local structures will map onto other hubs in the network, all of which must be similarly structured.

The meta-shift that had occurred through the process of developing and elaborating our landscape—in effect, a movement from "excavating" or "uncovering" implicit connections to "creating" or "inventing" new categories and structures—was not lost on participants. There was a clear sense that we were working at the edges of current cultural understandings, evidenced by such questions as "Has anyone done this before?" and "Does anybody else know about this?" For me, such queries are indicative of a critical feature of contemporary work with teachers: its participatory nature. This manner of research engages teachers as participants in the articulation and development of new knowledge, in effect repositioning them as agents in establishing knowledge rather than conduits of established knowledge—a point that I revisit in my closing remarks.

Further to this point, not only did the landscape activity contribute to an awareness of participation, it opened the possibility that teachers and their students might actually have a disproportionate influence on disciplinary mathematics as they select and connect (whether deliberately or accidentally) images, metaphors, applications, and exemplars that frame collective understandings of mathematical concepts. Far from merely representing what is already known, we recognized ourselves as participating in defining possibilities for knowers.

COMPLEX KNOWERS

In presenting the tension between hierarchical and networked conceptions of the structure of mathematics, I do not mean to suggest an irresolvable dichotomy. On the contrary, the dyad seems to be suggestive a more subtle tension that must be negotiated by mathematics teachers every day—namely the logical structure of collective mathematics knowledge and the analogical character of individual mathematics knowers.

A hallmark of mathematics is that, for a claim to be considered valid within the domain, it must be demonstrated to be logically consistent with established claims. Unfortunately, too often it seems that this criterion of logical validation is transmuted into a prescription for instruction, in essence leading to teaching emphases that are oblivious to the fact that humans are not logical beings. Rather, we are analogical (association-making) creatures whose capacity for logic rides atop our predilection to make connections. In other words, in the ongoing work with teachers, it has become more than obvious that any attempt to grapple with mathematics *knowing* is simultaneously a matter of grappling with the natures of mathematics *knowledge* and mathematics *knowers*. To this end, most of our gatherings have had a reflective component in which we "step outside" our discussion of mathematics in order to examine the structure of our engagement. How was it organized? How did we work together? When did things fall apart? Why? How might this matter for the classroom? In the process, we have been acutely aware of how our mathematics knowing is situated, distributed, and dynamic.

To elaborate, an early realization was that it was important for us to meet in actual classrooms rather than staffrooms or boardrooms. Having engaged with many groups with the "What's multiplication?" activity, for example, I can attest that the immediate presence of number lines, multiplication tables, and other pedagogical artifacts, not to mention access to texts and the Internet, has a dramatic effect on collective insight. More than mere representations of intelligence, such artifacts can serve as reminders of old and triggers for new insights—in effect, they are bestowers of intelligence.

A closely related, but somewhat more stimulating realization in our shared work is the distributed character of our knowledge of mathematics. As illustrated by both the list and the landscape, above, the understandings of multiplication that have arisen in this group are not simple representations of what every individual knew, but composite creations of what we are collectively coming to know. More profoundly, the "knower" in this case is not the individual, but the group. On this matter, we have been quite conscious and deliberate about how questions are posed, how interactions are structured, and how insights are represented, paying particularly close attention to participants' commonalities and diversities, as well as needs for independent and interdependent action (see Davis & Simmt, 2003). To this end, we have come to refer to our collective inquiries as "concept studies," which we define (in open terms) as collective learning structures through which we identify, interpret, interrogate, invent, and elaborate images, metaphors, analogies, examples, exemplars, exercises, gestures, applications, and so on that are invoked, sometimes explicitly and sometimes implicitly, in efforts to support mathematical understandings. In other words, they are conscious and deliberate efforts to work together to understand and apply precisely the sorts of ideas presented in this book.

The cohort of middle school teachers mentioned earlier invented the phrase "concept study" about one year into our shared work when a participant in the group suggested it would be useful to name what it was we were doing. We quickly homed in on the phrase, borrowing from two prominent strands in contemporary mathematics education research: first, the mathematical emphases of "concept analysis" (e.g., Leinhardt, Putnam, & Hattrup, 1992); second, the collaborative structures of "lesson study" (e.g., Fernandez & Yoshida, 2004). Concept analysis, which was particularly prominent in mathematics education research from the 1960s to the 1980s, has been focused on explicating logical structures and associations of mathematical concepts. The more recent lesson study movement is a collective-oriented structure for the articulation, critique, and further development of mathematics teaching strategies. To these we add a few other important ingredients in the concept study mix, including emergent research into the embodied nature and figurative dimensions of mathematical concepts and the situated and distributed character of mathematics knowing.

ENGAGING TEACHERS?

As a person with one foot strongly planted in mathematics education research community and the other placed among groups of practicing teachers, I am often asked by research colleagues how practitioners are thinking about insights into mathematics learning and knowledge that are trickling in from cognitive science, linguistics, sociology, cultural studies, and other domains. My answer is always that, with few exceptions, teachers embrace the ideas enthusiastically and intelligently—as long as they are engaged as co-producers of rather than recipients of knowledge. After all, how can issues of embodied knowing or distributed intelligence be engaged in ways that are not themselves embodied and distributed? This point is not a minor one. It is, in fact, a commentary on the relationship between researchers and practitioners—which, as illustrated by the introduction of radical constructivism some 25 years ago, has not always been a mutually productive one. I was a middle school mathematics teacher in the 1980s when details of this "ground-breaking" and "earth-shaking" theory began to arrive in dribbles and drabs, mentioned in keynote addresses at teachers' conventions, tucked into abstracts of annual conferences, and highlighted in rationale statements for new curriculum initiatives.

Unfortunately, this manner of introduction was incoherent at best. There was always a sense of something important, something I should know, something I should be doing, but the what's and the why's never seemed to be well connected. I learned I shouldn't teach *for* or *about*, but *through* problem solving, apparently because learners must be allowed to make their own sense. I learned that curriculum events should be multimodal and, in particular, that students should have ample opportunity to manipulate objects before leaping to abstractions, which had something to do with the fact that understandings are shaped and constrained by experience. I learned that students should be invited to re-symbolize their sense making in discussion and in writing, since externalized representation was good, but it seemed that internal representation was a grand misconception.

So I did my best to do all these things. As it turned out, however, neither my thinking nor my practice changed much. I make no claims to being particularly enlightened, but like most of the Western world I knew that we all constructed our own understandings. Thus my students were already engaging in diverse activities. I was well aware that knowing was conditioned by experience, context, support, motivation, and so on, and so they were already being asked to explain, defend, and elaborate their thinking. What was new here?

Of course, some years later when I finally had an opportunity to learn about what was really "radical" about radical constructivism, I was quite taken aback. I began to realize that I had not really understood what was being suggested or where it was coming from. Part of the reason was that, as with most of my teaching colleagues, my hearing was impaired by anxieties around covering the curriculum and preparing for examinations. Another aspect, I suspect, was that many of the champions of "constructivist teaching" that I encountered really did not appreciate (or at least failed to communicate) the subtleties and complexity of the discourse, opting for sound bites and pithy advice rather than engaging with the issues and tensions that were presented. But the major obstacle was that I simply did not have a good sense of what it was I took-for-true about knowing, learning, and teaching—and without some means of engaging intensively and extensively with these matters, there was not much hope of deep insight while dealing with the demands of classroom life.

In my current position as an educational researcher and teacher educator, I find myself positioned differently in relation to, but no less frustrated by, questions of what to do about emergent insights into knowledge and learning. In fact, the situation is amplified. Contemporary work on embodied and implicit knowing is simultaneously more compelling to me as a researcher and more challenging as a practitioner. How might teachers be involved in the conversation in ways that enable them to do more than implement new emphases and echo catchy, but assimilable sound bites? More specifically, how can we honor teacher knowledge while critiquing it? The most effective way I know of is to be part of it. To that end, staying with the topic of multiplication, I highlight two more "products" of collaborative work with teachers before closing with a few thoughts on what I would argue to be vital elements for pre-service and in-service teacher education. Both of these products might be described as "meta-interpretations" of the concept of multiplication, in that they arose in teachers' collective efforts to think through how the concept of multiplication might be framed in ways that are attentive to both horizontal (i.e., in-grade) and vertical (i.e., across-grade) connections and elaborations. The first is more focused on algorithms, while the second is much more conceptual in nature.

The first came up in the first concept study group that Elaine Simmt and I worked with, mentioned at the start of this chapter. That particular cohort of 25 had teachers from all grade levels, and so we took a particular interest in how multiplication might be represented in a manner that stretched across the school experience. How might, for example, the fourth-grade teacher frame multiplication of multi-digit whole numbers in a manner that extends what was learned in earlier years and anticipates what will be encountered in later years? It would be a serious misrepresentation to suggest that the discussion of this issue was a friendly and immediately productive one. On the contrary, it was heated and uncomfortable—crammed with accusations of ignoring one another's pedagogical insights, of slavish obedience to entrenched procedures, and of lack of awareness of the damage done by shortsighted curriculum choices. Nevertheless, the group persevered and, while never reaching consensus, did manage to find a space where all could agree on the utility of a specific interpretation—namely a grid-based algorithm. Not only does this meta-interpretation connected several images, metaphors, and processes of multiplication (in particular the discrete multiplication-as-array-making and the continuous-multiplication-as-area-making, as illustrated in Figure 9.2), it also highlights the similar processes at work in multiplication of whole numbers, decimal fractions, mixed numbers, and binomial expressions (among other applications).

The point here is not at all that multiplicative processes *should* be taught in grid format. It is not a prescription; rather, grid-based multiplication is a blend of interpretations that serves both to connect emphases across grades and to foreground figurative associations that might otherwise fade into tacit assumption. It is simultaneously a tool to link and a means to remind, intended more *for the teacher* than the student. (On the matter of pedagogical utility, however, it does have some distinct advantages over other algorithms, including clarity of process, built-in estimate, and built-in check.)

The second meta-interpretation, developed only over the past few months by the cohort of middle school teachers, is presented in Figure 9.3. It represents a rather complicated blend of interpretations, including multiplication as number line stretching and multiplication as a one-to-one mapping function, both of which have particular relevance to concepts and applications that arise in the middle grades. An exciting aspect of this particular blend was the "appearance" of stretched and compressed number lines on the xy-plane when vertical lines are constructed (see the central image of Figure 9.3). But the really exciting feature of this blended meta-interpretation popped out when the x-axis was extended in the negative direction and lines with negative slopes were added (depicted in the right-hand image of Figure 9.3). With those moves, it suddenly became obvious to participants that the product of two negative values will map onto the positive y-axis. In the process, a new image for multiplication by negatives emerged for the group—namely, as number line inversion as it compresses and passes through the origin. (This interpretation is quite distinct from the more commonly mentioned number "line rotation," particularly in terms of the associated action of compression through the origin, as contrasted with rotation about zero.)

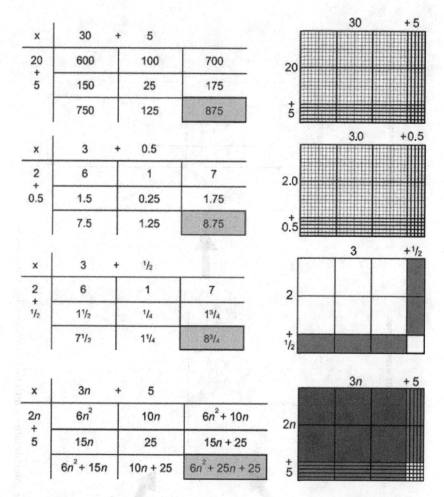

Figure 9.2.   A grid-based interpretation of multiplication, applied to whole numbers, decimal fractions, mixed numbers, and binomials.

As with the previous meta-interpretation, the suggestion is not at all that this composite image should become part of the grade school mathematics curriculum. Quite the contrary, it would likely be the source of much confusion. The point is, rather, that for this particular group of teachers, its creation served as a powerful moment of shared insight. It was an opportunity to link and elaborate knowings. It was an instance of participating in the production of teachers' disciplinary knowledge of mathematics.

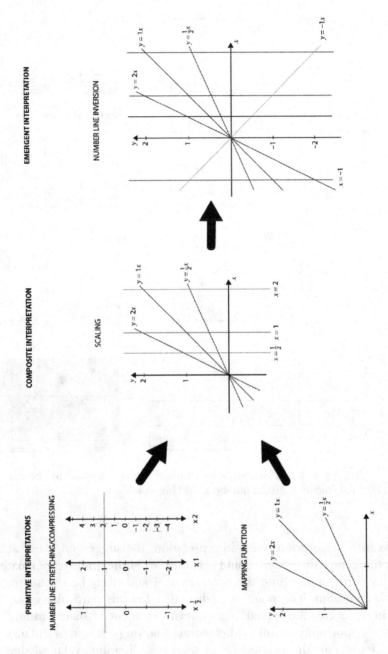

Figure 9.3. Blending figurative interpretations into more complex interpretations.

SOME CLOSING THOUGHTS

Before closing, I should explain that although I have focused on only one concept in this chapter, in fact various concept study groups are exploring a range of curriculum topics, including zero, one, equality, exponentiation, functions, problem solving, algebra, and algorithms. In this writing, I focus on multiplication principally because of its centrality to the K–12 mathematics curriculum. It is arguably the most significant of the major hubs in the network of school mathematics. I also chose it because, for me, it is a powerful illustration of what can happen when practicing teachers are invited into emergent research as co-producers rather than consumers of research insights. To that end, with regard to structures of collective engagement, some points of emphasis that have emerged across the cohorts with which I have worked include:

- opening definitions—excavating the images, metaphors, applications, and exemplars that are used to frame the concept;
- exploring implications—getting a sense of the local network of associations around the concept by examining nearby nodes, hubs, links, and entailments;
- mapping out a landscape—identifying horizontal and vertical curriculum connections while looking for similarities and differences among interpretations; and
- inventing meta-interpretations (blending interpretations into more encompassing models, and subsequently examining those models for limitations and possible entailments).[2]

This list is clearly not complete. Nor is it some sort of recommendation for action. It is, rather, an indication of the sorts of activities that can be, under the right circumstances, of compelling interest to practicing teachers.

As for the impact on practice, we have yet to delve deeply into that question. In fact, we're only just now reaching the place where we can begin to examine the complexities of this issue. However, there are a few effects that can be reported. On the level of individual practice, for example, concept study participants consistently report on the emergence of a particular sort of ethos within their classrooms, one of deeper engagement, more probing questions, and greater pleasure. In our discussions, they attribute this emergence to their own repositionings in relation to mathematics, which in turn comes to define classroom engagements. Most of the 55 teachers currently involved in different cohorts now see themselves as active and capable inquirers, and all note that they enjoy mathematics considerably more than they have in recent memory—prompting me to re-emphasize my conviction that a critical element in teachers' disciplinary knowledge of mathematics is their attitude toward knowledge.

More globally, the products of this research have already begun to spill out past the confines of the various cohorts. Grid-based multiplication has now worked its way into the formal curriculum in the province of Alberta, where the first concept study group was located. Although this evolution cannot be directly attributed to that cohort, there is clear evidence that reportings (e.g., at mathematics teachers' conferences, and to Ministry officials) of that shared work were highly influential. Further evidence of the influence is that I rarely get a simple answer "What's multiplication?" when I ask Alberta teachers—whereas I used to be able to count on it (and still can in most contexts). Word of the concept's broad network of associations has spread considerably in the last few years.

Once again, I am convinced that the critical element in this enterprise of presenting teachers with emergent insights into mathematics knowing is not a matter of finding ways to *introduce* the ideas, but of finding ways to *involve* teachers in the elaboration of the ideas. Such elaboration is not a mere matter of identifying implications. Rather, I would argue that it is nothing less than a reconfiguration of what it means to teachers' relationships to mathematics. Spurred by the realization that children and their teachers contribute to cultural mathematics through the selection and development of images, metaphors, analogies, gestures, and exemplars that come to be woven into conceptual understandings, teachers have to be active participants in this work.

NOTES

1. Although exceeding the intentions of this chapter, the significance of the highlighted elements in this "landscape" are worth noting. As developed in Davis (2008), some interesting tensions (and even contradictions) begin to arise when one explores the entailments of different images and metaphors for such concepts as product, factor, commutativity, and prime.

2. For more on these issues see Davis (2008), and Davis and Simmt (2006).

REFERENCES

Ball, D. L., Hill, H. C., & Bass, H. (2005, Fall). Knowing mathematics for teaching: who know mathematics well enough to teach third grade, and how can we decide? *American Educator*, 14–17, 20–22, 43–46.

Chevellard, Y. (1991). *La transposition didactique: Du savoir au savoir enseigné*. Paris: La Pensée Sauvage.

Davis, B. (2008). Is 1 prime? Developing teacher knowledge through concept study. *Mathematics Teaching in the Middle School, 14*, 86–91.

Davis, B., & Simmt, E. (2003). Understanding learning systems: Mathematics teaching and complexity science. *Journal for Research in Mathematics Education, 34*, 137–167.

Davis, B., & Simmt, E. (2006). Mathematics-for-teaching: an ongoing investigation of the mathematics that teachers (need to) know. *Educational Studies in Mathematics, 61*, 293–319.

Fernandez, C., & Yoshida, M. 2004. *Lesson study: A Japanese approach to improving mathematics teaching and learning*. Mahwah, NJ: Lawrence Erlbaum Associates.

Leinhardt, G., Putnam, R., & Hattrup, R. A. (1992). *Analysis of arithmetic for mathematics teaching*. Hillsdale, NJ: Lawrence Erlbaum Associates.

Mowat, E. (2008). Making connections: Mathematical understanding and network theory. *For the Learning of Mathematics, 29*(3), 20–27.

CHAPTER 10

EXPRESSIVENESS AND MATHEMATICS LEARNING

Ian Whitacre, Charles Hohensee, and Ricardo Nemirovsky

How do multiple representations come together and participate in mathematics learning and insight? This is the broad question that we explore in this chapter. We observed the activities of students in an environment of multiple representations. The primary phenomenon that we examine here is how aspects of these multiple representations became incorporated into streams of bodily activity, exhibiting features of expression. As Noble, Nemirovsky, Wright, and Tierney (2001) note, there appears to be a general view "that students connect experiences in different environments by recognizing a core mathematical structure that is common to all environments" (p. 85). These authors challenge this view and instead "propose an alternative perspective on learning, in which students make mathematical environments into lived-in spaces for themselves and connect environments through the development of family resemblances across their experiences" (p. 85). In this chapter, we examine a related but different research question: not how different lived-in spaces interrelate through family resemblances, but how different representations simultaneously intermingle within a certain lived-in space. The representations we consider include a force sensor, a car moving along a track (its acceleration depending on

---

Mathematical Representation at the Interface of Body and Culture, pp. 275–308

the force applied to the sensor), and a real-time graph of force/acceleration versus time on a computer and projector screen.

We present two 45-second episodes from a teaching experiment, focusing on a particular student in each episode (Maria, then Crystal). In the first episode, Maria describes her understanding of how to make the car stop. In the second episode, Crystal describes how she imagines the car will move according to a given graph. This case study illustrates how different representations entwine in the activities of the body in concert with attempts to regulate interpersonal transactions (e.g. moments of introspection or references to past conversations). Immersed in the students' bodily activities, the representations do not preserve well-defined boundaries. Furthermore, the distinction between signified and signifier appears fluid and sometimes mutually reversed. We suggest that the mathematics entailed in the use of multiple representations is to be found in persons' multifaceted bodily engagement with them. Rather than locating the mathematics in objects or inscriptions, we suggest to place it in the lived-in spaces of the symbol users, a place in which various representations become infused in a stream of activities that cross over distinctions between signified and signifier.

INTRODUCTION

Mathematics educators often use the word *representations* in reference to notations or symbol systems that are used in mathematics. Common examples that are easily recognized as representations are Cartesian graphs, number tables, and equations. Different researchers have advanced the thesis that a major component of mathematics learning is the development of the ability to translate amongst different representations. Meira (1998) has accounted for the role of human activity by emphasizing that mathematics does not exist within the representations themselves but in their active use. Hiebert and Carpenter (1992) promoted the distinction between observable external representations and unobservable internal representations and have argued that the external and internal are related in important ways.

Implicit or explicit in the varied accounts of and perspectives on representations is that they are signifiers whose meaning—what they signify—is to be found outside of themselves. In this view, a graph of a function is not the function itself, but merely one of various possible ways of representing it. Representations are usually taken to suggest that (a) the relationship between them can be completely described in terms of necessary conditions (e.g. the relationship between a linear graph on a Cartesian plane and the corresponding linear equation), (b) the represented is determined by means of taxonomies and class definitions (e.g., linear equations with positive slope and a graphical lines slanted up both repre-sent a certain subset of linear functions), and (c) representations can be translated with preservation of content (e.g., the same set of numbers can be represented in many ways, all of them subject to mutual translation).

In this chapter, we distinguish between representation and expression and claim that what symbol users do with mathematical representations can be better described as incorporating representations into bodily expressions, rather than as translating between them or organizing them by taxonomies and necessary conditions. The distinction between repre-sentation and expression became a centerpiece of philosophy of art during the twentieth century. Whereas it had long been understood that the value of a piece of art was not confined to its representational power (e.g. the extent to which a painting of a house resembles that house) but that it encompassed a multitude of expressive aspects (e.g. the house as a nurtur-ing or threatening place), it was twentieth-century art that brought the notions of representation and expression to a vivid collision. The paintings of Jackson Pollock, for instance, are not representative of things in the world or even in some fictional worlds. The meaning of these works does not derive from things that are represented (i.e., located outside of the paintings themselves) but from within—from the particular details of the artist's use of color and texture and the shape of the strokes or drizzles of paint applied to the canvas. All of these elements taken together express something, much in the same way that a piece of instrumental music is expressive. They defy analysis in terms of representation, not only because nothing can be plainly and without question pointed out as represented by them, but also because there are not means of translating them with pres-ervation of content, of organizing them according to formal taxonomies, or of making them compliant with necessary conditions. In Table 10.1 we summarize the distinction between representation and expression.

Our central thesis is that symbol users incorporate mathematical repre-sentations into bodily expressions. By *incorporating*, we mean the merging of multiple representational aspects through the flow of several strands of bodily activity—voice, gesture, gaze, and posture—constituting the enact-ment of mathematical expressions. We study the use of multiple represen-tations by investigating the lived-in spaces of the symbol users. A lived-in space refers to the place someone is dwelling in—the subjective world that one experiences at a given time and location, being understood that, for our purposes, subjective worlds are the real ones.

REPRESENTATION VERSUS EXPRESSION

Philosophers of art have discussed both the representational and expres-sive qualities of artwork. Since humans began to make cave paintings in

Table 10.1. Summary of the Distinction Between Representation and Expression

Representation	*Expression*
The represented is subject to complete description	The expressed is amenable only to partial and hard-to-pin-down descriptions (e.g., that which is expressed by a piece of instrumental music)
The represented is transcendent to the representation (e.g., the represented is to be found outside the representation, such as a city that is outside of its map)	The expressed is immanent to the expression (i.e., we get the sense of an expression by immersing ourselves in it, rather than by shifting to a signified realm located elsewhere)
The represented is determined by means of taxonomies and class definitions	The expression is determined by the unique particulars of the expression (i.e., no detail of an expression is irrelevant to what it expresses)
The representation stands for the represented	The expression is part of the expressed (e.g., facial expressions of grief or surprise are part of these emotional experiences)
Representations are translatable	Expressions are untranslatable (e.g., a sculpture may inspire a piece of music, but it does not get translated into music)

prehistoric times, there has tended to be a representational aspect to visual art. Children draw pictures of things in the world—their selves and family members, the sun, a tree, a house. They might also draw things that are imaginary—dragons or fairies, for example. In either case, these drawings are representations *of* something. There is the house, and then there is the child's drawing of the house. The drawing is meant to stand for the house and be understood as such, without being confused for the genuine article. Representation can be defined in terms of a correspondence between a representing world and a represented world (Kaput, 1987). In this case, the child's drawing is the representing world, and the house and surrounding scenery constitute the represented world. Entities in the drawing correspond to objects in the vicinity of the house. Such a representative relationship is unidirectional: the drawing represents the house, and not the other way around. Many works of visual art, as well other art forms, exhibit this representative quality.

Benedetto Croce is known for impacting philosophy of art by suggesting that expression, rather than representation, should be the sole criterion for the aesthetic value of art. In the Crocean view (Kemp, 2003), the accuracy of representation is irrelevant. Representation smacks of imitation. The aesthetic meaning of art derives not from what it represents, but from what it expresses. This expressive quality manifests from within the work itself, rather than borrowing meaning from some outside world. Painters like Jackson Pollock shook up the art world in the twentieth-century by creating non-representative works. Pollock is famous for his "drip" technique of spontaneously spattering liquid paint onto canvas. His works defy categorization in terms of representation. We cannot identify in them any object being represented. There is only the paint on the canvas and the expression that it conveys. This style of art earned Pollock and others the designation *abstract expressionist*.

Most works of art can be viewed in terms of both representation and expression. They may contain elements that correspond to entities in a represented world. At the same time, however, their meaning is not summed up by a list of correspondences between worlds. Additional meaning is conveyed via the technique and composition employed. The paintings of Jean-Michel Basquiat (see Figure 10.1) exemplify artwork that includes representative elements but defies categorization as primarily representational. In contrast to, say, a realistic landscape, it can be difficult to identify a background, middle ground, and foreground in Basquiat's paintings. They have a collage-like quality, juxtaposing representative and non-representative elements within the same expressive whole, which creates ambiguities (e.g., between figure and ground). Certain elements appear representative of objects, real or imaginary. Others deny any such correspondence and yet contribute undeniably to the expressive quality of the composition. As Basquiat is widely quoted as having said, "Every single line means something."

Figure 10.1.   *Untitled,* Jean-Michel Basquiat (1981) .

BACKGROUND OF THE STUDY

We report on a teaching experiment in which participants interacted with a device that connected three representations of acceleration:

1. The force sensor: The "forcer" (the person applying force to the sensor) did so by pushing or pulling against it with two fingers (typically, the index and middle fingers of his or her dominant hand).

2. The car and track, particularly the movement of the car along the track: The participants experienced the car's movement by following it with their eyes, as well as sometimes feeling the car with a hand as it moved.

3. Real-time graphs of acceleration/force over time: The participants watched the graphs as they were created and also considered completed graphs. The graphs could be regarded as describing force (applied to the sensor) and/or acceleration (of the car) over time.

In keeping with mathematics education research literature, we use the term *representations* to refer to the force sensor, car, and graphs in the sense that each of them is amenable to description in terms of self-contained properties whose attributes can be mapped out with external properties (e.g., increasing positive force applied to the sensor correlates with soaring height on the graph). However, as these three representations were actually used by students, and as the students became skillful in their use, what emerged were bodily activities whose qualities strikingly resemble those of an expression, as described in the right column of Table 10.1. The ways in which students interacted bodily with each of these representations is fundamentally relevant to our analysis.

The episodes that comprise the data that will be presented here come from a teaching experiment conducted by the authors with four tenth-grade students during the spring of 2008. First, we describe our research methods, including our theoretical framework, the teaching experiment, and our methods of analysis. We then present transcripts of two episodes, followed by our analyses of the observed behavior. Finally, we discuss the roles of representation and expression in students' experiences in an environment of multiple representations.

The theoretical framework undergirding the following analyses derives from the construct of lived-in spaces (Nemirovsky, Tierney, & Wright, 1998). As stated earlier, a lived-in space refers to the place someone is dwelling in—the subjective world that one experiences at a given time and location. Even though each individual dwells in an ongoing distinctive lived-in space, the study of lived-in spaces is not amenable to being treated as strictly individual for the following reasons: (a) each lived-in space is co-constituted by the participant, the others, the symbols, the lay-out, and the objects present in the place; (b) whatever traits we find or observe belong to the place the participant lives in (they are not features that the individual carries as a characteristic, say, of her personality); and (c) any lived-in space is ephemeral, unstable, and subject to drastic trans-formations from moment to moment. We examine lived-in spaces by trac-ing and interpreting acts of symbol-use as bodily activities. The intimate relationship between lived-in spaces and bodily activity stems from a core thesis of embodied cognition:

> If it is true that I am conscious of my body via the world, that [my body] is the unperceived term in the centre of the world toward which all objects turn their face, it is true for the same reason that my body is the pivot of the world ... I am conscious of the world through the medium of the body. (Merleau-Ponty, 1989, p. 82)

Bodily activities span a complex array of parallel threads, which at times join altogether and at other times seem to take on related but different matters of concern. Bodily activities of different participants often engage in a striking choreography indicative of how the participants' lived-in spaces shape one other. We study bodily activities by examining utter-ances. For us, utterances encompass all aspects of bodily activity that play a part in a given conversational turn, including facial expression, gesture, tone of voice, sound production, eye motion, body poise, gaze, and so forth. Here we describe four particular features of students' utterances, as seen in the two episodes. In our view, these features reflect the intermingling of representations within students' bodily expressions. These four features are *Convergence*, *Participation*, *Transitional Smoothness*, and *Interpersonal Transactions*.

First, by *convergence* we denote the fact that a single utterance may simultaneously integrate aspects of different representations. We exam-ine, for instance, how a single motion of the arm/hand expresses at once an action on the force sensor and a motion of the car. Second, certain qualities emphasized by a student may permeate all aspects of an utter-ance. For example, a student describes the motion of the car, which she anticipates to be slow, by saying the word *slowly*, while moving her hand slowly, talking slowly, and adopting a relaxed facial expression. None of these aspects stand apart as signifiers, but they constitute an expressive whole similarly to how a facial expression of sadness does not signify sad-ness but is part of being sad. Through *participation*, utterances transform the body into a field of expression. Third, one of the essential features of an expression is that its meaning (to the observer) emerges from within. To appreciate a piece of instrumental music, for instance, it may well be valuable to learn about its history and its relationships to musical styles.

The bottom line, however, is what the piece expresses to the listener in the listening itself. Philosophers use the term *immanence* to describe this power of meaning from within. We propose that smooth transitions between utterances reflect immanence—hence our concept of *transitional smoothness*. It is not necessarily that an utterance ends and a new one begins, enacting a break that signals the transition from one to the other. At times, an utterance smoothly transforms itself into a new one without discernible breaks. For example, we see transitional smoothness in the second episode when Crystal transitions from describing the speed of the car to indicating its direction. It is not that she suspends the expression of slowness and then resets her body to indicate a direction of motion; instead, her right hand keeps moving slowly as she gradually starts to fold her fingers, leaving only the index finger extended, as her arm shifts upwards, and her voice begins to produce the corresponding verbalization indicating direction. Crystal's utterance does not shift from one discrete state to another. Rather, it emerges organically, reflecting how the direction of motion was already immanent in her expression of slowness.

Fourth, punctuations between utterances are often signaled by moves regulating *interpersonal transactions*. Crucial aspects of bodily expressions that incorporate graphical shapes, movements, and forces, are those that reflect intentions: to be understood, to help someone else perform, to request an explanation, and so forth. What is being expressed is not merely information regarding the explicit topic (e.g., how the car will move), but at the same time an underlying layer of communication concerning expectations for the interaction. For example, a student makes an utterance describing the motion of a car. Then her abrupt withdrawal of her hand signals to her interlocutor that he is free to begin his utterance. In this way, gestures convey meanings that help facilitate the social interactions with others in the lived-in space.

TEACHING EXPERIMENT

Here we present an analysis of selected episodes from activities conducted with a small group of tenth-grade students over four one-hour sessions. The goal of these activities was for the students to learn about acceleration. In spite of acceleration being a pervasive aspect of everyday life, physics and mathematics teachers often face serious difficulties in helping their students become fluent with the science and mathematics involved in understanding acceleration. One component of these difficulties is perceptual: studies have shown that humans cannot visually discriminate acceleration very easily, nor sense it bodily apart from velocity, tilt, and other aspects of our motions.

We explored the use of an experimental tool designed to facilitate learning of acceleration. The experimental tool used in this study allowed students to exert forces on a force sensor, which determined the acceleration of a metal car moving along a track in real time. A computer projector simultaneously displayed a graph of force/acceleration over time, which could be replayed together with the corresponding motion of the car. The hypothesis of the device designers is that such visuo-motor correlation between force and acceleration might help students develop a richer sensitivity to variations of acceleration. We present two episodes from the teaching experiment.[1]

Episode 1: Maria

Synopsis

The following 45-second episode takes place near the end of Session 3. Two students, Maria and Nhu, participated in this session. It is important to understand the context in which the utterances that follow took place. Toward the end of the session, Nhu was operating the force sensor and attempting to produce a stop. In the run discussed in this episode, the car traveled backward, as intended, but failed to stop. The run is replayed, and Ricardo asks Nhu why she thinks the car failed to stop. Maria answers, offering an explanation related to "balance," which she elaborates on in response to follow-up questions from Ricardo (16) and Ian (18).

Transcript

1   Ricardo: Okay. It starts the:re. So, then you pu:ll. And you wanted it to stop, but it kept going. Why did it keep going?

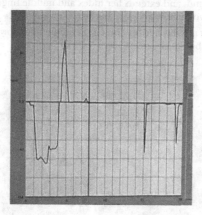

Figure 10.2. (00:10.94)

2   Maria: Oh, because you didn't put too much force_to go_

             ((Maria sticks out the index and middle fingers of her right hand. Her
             left hand rests on her hip. Her gaze is directed at the screen.))

3   ((Maria quickly retracts her middle finger and turns her hand in the
             forward direction of the track. She points in the forward direction with
             her index finger.))

4   Maria: _um_

             ((Maria sticks out her index and middle fingers again. She draws her
             right hand across her torso as she says "um." Her eyes appear either
             closed or looking down at the track.))

5   Maria: _forward._

             ((Maria quickly thrusts her right hand in the forward direction of the
             track. Her eyes appear closed or looking down at the track.))

Figure 10.3. (00:15.88) – "um"   Figure 10.4. (00:16.01) – "forward"

6   ((Maria's right arm swings down and is retracted behind her back. Her
             left arm remains fixed. Maria's head is pointed downward and her eyes
             appear closed.))

7   ((Following Maria's utterance in (5), Nhu looks off to her left, raises her
             left hand, and extends her index and middle fingers. She then retracts
             her two fingers, and puts her hand down.))

8   ((Nhu retracts her two fingers, bending them downward, while pulling
             her elbow backward.))

Figure 10.5. (00:18.28)   Figure 10.6. (00:18.55)

9 Maria: So:::o

((Maria twists her body away from the other participants and toward the track. She lifts her left arm off of her side and swings her left hand across her torso. At the same time, she swings her right arm out from behind her back.))

10   ((Maria positions both of her hands out in front of her. She looks down at her hands. Her elbows are raised from her sides.))

11 Maria: like I to:ld- I said before

((Maria's left hand points at Ian. Her right hand remains fixed. She smiles slightly.))

Figure 10.7. (00:19.81) – "I to:ld"

12 Maria: theres needs to be like some kind of balance

((Maria moves both of her hands, alternating (left and right) up and down, while her head points downward. Her eyes may be focused on her hands or perhaps not focused at all.))

Figure 10.8. (00:22.35) – "some kind of balance"

13 Maria: between the – um-

((Maria pauses, looks down, and closes her eyes. She strokes her face with her left hand as she says "um." Maria's right hand remains positioned as it was at the end of (12).)

14   ((Maria's left hand rejoins her right in the ready, public position out in front of her.))

15   Maria: the push and the pull.

                ((Maria releases her arms from the fixed position in front of her, allowing the left to swing freely. The right drops slightly but remains close to her waist.))

16   Ricardo: M::m. So, it stops when there is a balance?

17   Maria: Uh-huh.

                ((Maria's head quickly turns in the direction of the screen just as Ricardo finishes saying "balance".))

     Ricardo: [A push and a pull balance.

18   Ian: So what does she need to do (---) Maria?

19   Maria: Probly put more balance.

                ((Maria makes fists and moves them horizontally, alternating (left and right) near and far, rotating at the elbows.))

Figure 10.9. (00:36.43) – "put more balance"

20   Maria: If like-

                ((Maria extends the middle and index fingers of her right hand. She quickly retracts her arm slightly, moving from the shoulder.))

21   Maria   she pushes hard

                ((Maria's right arm swings down to her side. She raises her left hand and strokes her face.))

22   Maria: she needs

                ((Maria's hands return to the ready position and briefly move in alternating fashion, but only subtly.))

23   Maria: to pull – m – hard.

                ((Maria allows both of her hands to drop. Her arms swing back and forth. Her eyes are either closed or looking downward.))

24   Maria: Like, well, no::t–

                ((Maria makes another alternating gesture, this time on a diagonal plane. The fulcrum is at her wrists, rather than her elbows. Her hands are open, and she moves them up toward her torso and then down and away.))

Figure 10.10.

25   Maria: She needs to have a balance.
((Maria moves both of her hands, alternating (left and right) up and down, while her head points downward. Her eyes may be focused on her hands or perhaps not focused at all.))

Figure 10.11. (00:43.77) – "She needs to have a balance"

26   ((Maria strokes the right side of her face with her left hand.))

Analysis

We describe four features of Maria's activity and give specific examples. These features should be understood in terms of how they manifest in her bodily activity. Each feature speaks to the incorporation of representations into Maria's expressive acts.

Convergence. Maria's observed activity suggests that, at least at times, different representations were not regarded as separate entities. Rather, they seemed to converge and be treated more closely—more like a single entity. For example, Maria gestures an action on the force sensor with her right hand while simultaneously gesturing the movement of the car with the same hand and arm and saying "forward" (in turn 5). In interacting with the force sensor, Maria's experience of "forward" related to pushing forward with the index and middle fingers of her right hand. In observing the movement of the car along the track, "forward" related to the direction of the car's motion. While Maria's verbalization distinguished between applying force and moving forward, her gesture expressed both simultaneously. Students' experiences with the sensor, graph, and car had often been simultaneous. The students' activity suggests that these three representations were not regarded separately. On the contrary, the students negotiated and interpreted the various representations simultaneously.

2 M: Oh, because you didn't put too much force <u>to go</u>	M sticks out the index and middle fingers of her right hand. The two fingers are spread apart at approximately the width necessary to operate on the force sensor. Her left hand rests on her hip. Her gaze is directed at the screen.	 Figure 10.12. (00:15.28)
3	M quickly retracts her middle finger and turns her hand in the forward direction of the track. She points forward with her index finger.	 Figure 10.13. (00:15.48)
4 M: <u>um</u>	M sticks out her index and middle fingers again. She draws her right hand across her torso as she says "um." Her eyes appear either closed or looking down on the track.	 Figure 10.14. (00:15.88)

5

M: <u>forward</u>

M quickly thrusts her hand in the forward direction of the track as she says
"forward." Her eyes appear either
closed or looking down on the track.

Figure 10.15. (00:16.01)

In turns 2–5 Maria says, "Oh, because you didn't put too much force to
go—um—forward." We interpret this verbalization to mean that Nhu did
not apply enough forward force to the sensor to cause the car to stop
going backward and begin going forward. The interpretation articulated
in the previous sentence makes explicit the distinction between the forcer
and the car, as did Ricardo's question (turn 1). Maria's answer, on the
other hand, made no such distinction. She simply said, "to go—um—forward." If Nhu had "put" more force, who or what would have gone forward? Nhu? The car? Both of them in different senses? We acknowledge
that people often speak in an abbreviated manner that does not explicit
make details such as who did what entirely . However, statements could be
abbreviated in any number of ways, so it seems to us that the way in which
statements are abbreviated is nonetheless significant. In particular, we
suggest that Maria's language is characteristic of individuals inhabiting a
lived-in-space in which they interact with representations. We believe that
in a very real sense it is meaningful in such a lived-in-space to talk about
both the car and the forcer as "you."

The convergence of signified and signifier evident in Maria's utterance
is similar to the ways that people talk when playing video games. The
player, via some form of remote control device, determines the action
(some movement, say) of a character in the game. The real-time connection between person and on-screen character facilitates a fusion
(Nemirovsky et al., 1998) between the two. In these contexts, players are
commonly heard to say such things as "*You* have to go down there" or
"*You*'re catching up" or "Jump!" The "you" is ambiguous in these utterances, but not in the sense that any confusion is created. Indeed, "you"
may refer to both the player and the character simultaneously or as if the
two were one and the same.

Participation. We noted repeated instances in which students' bodies
participated with representations in intimate and inextricable ways.
Maria's balancing gestures provide a prime example of the bodily participation that we observed in the sessions. For each instance in which
Maria used the word *balance,* her utterance included what we describe as

a balancing gesture. We identified three instances of balancing gestures in the brief episode (turns 12, 19, and 25). In what follows, we discuss the context and characteristics of Maria's utterances concerning balance in some detail. We then relate our observations to the role of expression.

To make sense of the discussion of balance and balancing, it will be useful to recall the context in which these utterances took place. Nhu had just produced a run in which the car moved backward. She intended to make it stop at a specified location along the track but was unsuccessful. (The car ran all the way to the end of the track, which then caused it to stop and terminated the run.) Maria offers an explanation for the behavior of the car as it relates to Nhu's actions on the force sensor.

The complete graph of force/acceleration over time for Nhu's run was displayed on the screen throughout the episode. The participants had seen the graph produced in real-time. They had then seen the run replayed, again in synchrony with the graph. The graph now remained as a static record of the run. Although the graph was static, Maria's gestures were dynamic. She did not position her hands and arms in such a way as to demonstrate equal-sized upward and downward peaks in order to represent a balance. Instead, her hands rapidly oscillated in alternating fashion, suggesting an active process of *balancing*, despite her repeated use of *balance* as a noun. Thus, while *balance* was used as a noun in her verbalizations, Maria's balancing gestures were verb-like, thereby contributing additional content to the utterances.

12	M moves both of her hands, alternating (left and right) up and down, while her head points downward. Her eyes may be focused on her hands or perhaps not focused at all.
Maria:	
theres needs to be, like, some kind of balance	

Figure 10.16. (00:22.35)

19	M makes fists and moves them horizontally, alternating (left and right) near and far, rotating at the elbows.
Maria:	
Probly put more balance.	

Figure 10.17. (00:36.43)

12

Maria:
She needs to have a
balance.

Maria moves both of her hands, alternating (left and right) up and down,
while her head points downward. Her
eyes may be focused on her hands or
perhaps not focused at all.

Figure 10.18. (00:43.77)

Maria's gestures in turns 12 and 25 were oriented in the vertical plane—the plane of the screen on which the graph was located. Both of these utterances referred to balance as a property of the run that was not explicitly connected to the forcer's actions. When asked what Nhu needed to *do* (turn 19), Maria's balancing gesture transformed. Her hands formed fists, and the plane of the movement shifted from verti-cal to horizontal as she said, "Probably put more balance." Only this verbalization involved an action associated with balancing. While the notion of putting balance is rather vague, Maria's associated gesture took place in the plane of the track and the force sensor and involved alternating pushing and pulling movements. In terms of these charac-teristics, the gesture overlapped with Maria's experience of operating on the force sensor and observing the movement of the car. Maria's ver-tical and horizontal balancing gestures expressed senses of balance and balancing in two distinct but interrelated ways. In terms of the graph as a static record, there needed to *be* a balance (equal areas bounded by the curve and the horizontal axis, for the upward and downward peaks) in order to achieve a stop. In terms of acting on the force sensor, Nhu needed to *put* more balance (equal cumulative pushing and pulling forces over time) in order to achieve a stop.

We see in this pair of related gestures evidence that Maria's reasoning and communication about balance were intimately connected with the physical experience of her body in the lived-in space. These gestures were not direct reenactments of physical interactions with the device (she had certainly never manipulated the graphs with her arms in a direct sense) or literal demonstrations of the actions and conditions necessary to achieve a stop. They were instead expressions of balance and balancing, features of which overlapped with the corresponding representations (the graph and force sensor, respectively). Both gestures centered on—that is, were bal-anced about—her own body, as opposed to some external object or arbi-trary location in space. They emanated from within Maria, rather than simply signifying something outside of herself. In this way, Maria (body and all) participated with the representations around her as she expressed her ideas regarding balance.

Transitional Smoothness. During the episodes, students' gestures often transitioned from one idea to another with remarkable smoothness. For example, in turns 25–26, Maria says, "Like well no::t– She needs to have a balance." During the utterance in turn 25, Maria begins an alternating gesture, which is similar in many ways to her balancing gestures. However, there are some important differences. Physically, this gesture is distinct from her balancing gestures in that the fulcra are her wrists, rather than her elbows, and that the gesture is oriented on a diagonal plane. This gesture also coincides with a verbalization that indicates negation. Maria apparently wanted to revise her previous explanation (turns 19–23), returning instead to the point that Nhu needed to *have* a balance (turn 26). Thus, the alternating motion in this case seems to indicate dissatisfaction with the previous explanation. In turns 25–26, Maria's alternating gesture transforms. By the time she says "have a balance" the plane has become vertical and the fulcra have migrated to her elbows. This transition occurs smoothly, which is to say that we cannot identify a point at which she stops performing the negation gesture (turn 25) and begins to enact the balancing gesture (turn 26).

Interpersonal Transactions. The preceding examples have focused on acts of overt communication, which were characterized by their expressive nature. We note that during and between these overt communicative acts, an interpersonal layer of activity served as a kind of punctuation that structured the communication. For example, in turn 11, Maria points to Ian as she says "like I to:ld – I said before" in preparation for giving her balance explanation. Earlier in the session, she had introduced the idea of balance in an exchange with Ian. She was now preparing to express the same idea, this time to the whole group. Her pointing gesture served to locate that earlier conversation, not according to its physical location in the room but to the person with whom the conversation had taken place. This way of locating a lived experience highlights an important distinction between a person's lived-in space and the physical space itself. In this instance, it seems that the specific physical location of the earlier conversation was not as significant an aspect of the interaction for Maria as was the identity of her interlocutor.

11	Maria's left hand points at Ian. Her right hand remains fixed. Maria smiles slightly.
Maria:	
like I to:ld- I said before	

Figure 10.19. (00:19.81)

Maria's smile suggests that in this moment she awakened a previous episodic feeling associated with the earlier conversation. Accordingly, the memory of a past event often emerges together with a bodily poise that was experienced during the past event. In this case, Maria was poised to perform the balancing gesture that she had performed earlier in the session in a conversation with Ian. In that conversation, she had first expressed the idea of the necessity of a balance in order to achieve a stop. This idea had been developed and introduced exclusively by Maria. The researchers had made no such suggestion at any point. Thus, Maria likely felt a sense of ownership for this idea, which coincided with positive feelings.

Episode 2: Crystal

Synopsis

In this episode, Crystal and Tuan had been shown a hypothetical graph. In terms of force, the graph represented a run in which the forcer exerted a brief pull on the sensor, followed by a brief push (see Figure 10.20). The students were asked to predict the resulting motion of the car along the track. Note that this was the first time that these students were shown a graph with one upward peak and one downward peak. Prior to this, they had investigated the motion of the car for graphs with a single peak.

Crystal explains that she thinks the car will first move slowly backwards. Then she continues her explanation by gesturing forward with her hand and saying that the car will go slowly after the second bump on the graph.

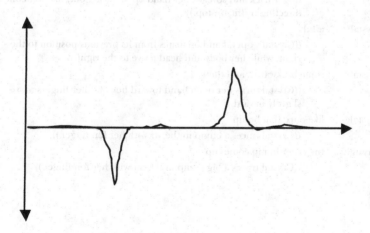

Figure 10.20.   Force-time graph that was shown to Crystal.

As she describes the car moving slowly, Crystal's hand is moving in the forward direction of the track. Charles asks her if she means that the car will go forward or that the car will continue to go backward. She clarifies her prediction that, after the second bump, the car will go forward.

Transcript

1   Crystal: I think that it's gonna

((Crystal rocks in her chair, leaning nearer to the track as if to position herself to look along the track.))

Figure 10.21.

2   Crystal: come backwards really slowly
((Crystal points to the far end of the track, then draws her pointing finger back and spreads her hand open, facing out, like someone directing traffic to stop.))

3   Crystal: w'l::l
((Crystal's open hand oscillates from its previous position to the right, while her body and head move to the right.))

4   Crystal: come backwards, and then
((Crystal turns her open hand toward her, and her fingers move slightly inward.))

5   Crystal: like once that bump
((Crystal traces a bump in the air with her forefinger.))

6. Crystal: the other bump comes up
((Crystal traces a big bump in the air with her forefinger.))

Figure 10.22.

7   Crystal: I think it's gonna go slo:::owly

((Crystal's hand assumes a claw shape and then flattens, shifting to a palm-down orientation. Her hand slides forward as she says "slo:::owly".))

Figure 10.23.

8   Crystal: then it's gonna keep that same pace and go slowly

((Crystal continues sliding her hand forward slowly. She then lifts it quickly as she completes her verbalization.))

9   Charles: Now you're moving your hand this way

((Charles imitates Crystal's gesture, moving a flat hand, palm-down, in the direction of the far end of the track. He looks toward that end of the track.))

Figure 10.24.

| 10 | Charles: | Do you mean it's moving that way? |

((Charles repeats his previous gesture, but turns to look at Crystal as he says "moving."))

| 11 | Charles: | Or do you mean moving this way? |

((Charles reverses his gesture so that the same hand moves toward the near end of the track. His gaze is on Crystal. He freezes at the end of gesture until Crystal says "goes."))

| 12 | Crystal: | Once it goes backwards |

((Crystal makes a first with extended thumb and points behind her with the thumb.))

| 13 | Crystal: | but then once you go forward |

((Crystal points forward with her forefinger.))

| 14 | Crystal: | once that bump go forward |

((Crystal maintains her pointing gesture while tracing a bump in the air with her forefinger.))

Figure 10.25.

15 Crystal: it moves slowly

((Crystal flattens her hand, palm-down.))

16 Crystal: that way

((Crystal's hand slides slowly forward and then makes a quick point-ing gesture at the end of the motion.))

Analysis

We describe four features of Crystal's activity and give specific exam-ples. As in the analysis of the first episode, each feature speaks to the role of representations in Crystal's expressive acts.

Participation. We see examples in this episode in which aspects of math-ematical representations get expressed in Crystal's bodily actions in ways that cannot be adequately described it terms of a signifier standing in for something signified. Instead, her bodily actions participate in expres-sions. In turn 7, Crystal explains what she thinks the car will do at the end of its run.

7

I think
it's gonna
go slo:::owly

Crystal's hand
assumes a
claw shape
and then
flattens,
shifting to a
palm-down
orientation.
Her hand
slides forward
as she says
"slo:::owly."))

Figure 10.26.   (00:13.3).

8

then, it's
gonna keep
that same
pace and go
slowly.

Crystal
continues
sliding her
hand forward
slowly. She
then lifts it
quickly as she
completes her
verbalization.

Figure 10.27.   (00:16.8).

Crystal illustrates her expectation for the motion of the car by moving her hand. As she moves her hand slowly forward, she looks at her hand as she says the word *slowly*. Here and throughout the rest of the episode, Crystal uses various bodily motions in communicating about the speed of the car. These bodily motions participate in her expressions of motion, in the same way that a smile participates in (rather than represents) happiness. Researchers have found correlations between people's emotions and their abilities to make associated facial expressions. Gestures such as Crystal's expression of the car's motion by embodying the car with her hand, appear to share this feature of participation.

The fact that Crystal looks at her own hand and slows her utterance down as she says "slowly" suggests that she is trying to synchronize her utterance with her hand. It also appears that her hand not only signifies the car's motion, but that she is describing the motion of her hand in its own right. Thus, the hand becomes a participant in the expression rather than just representing the car's motion. To an extent, the motion of her hand takes on a life of its own.

Crystal's gesture seems to be tied to her understanding of the motion of the car and not just representative of it. A similar effect is observed when protein crystallographers expressed their understanding of the structure of molecules with their bodies: "As they manipulate and build crystallographic models, they incorporate these complex molecular forms into the folds of their flesh. In the process they rearticulate and so entrain their embodied imaginations" (Myers, 2008). By studying her hand and synchronizing her utterance to her gesture, Crystal also seems to be entraining her embodied imagination to the kind of motion that would be appropriate.

Convergence. We also noted instances of convergence in Crystal's utterances during the episode. Just as artistic expressions can create ambiguities of meaning, combining gestures pertaining to mathematical representations can create ambiguities, especially to an outside observer. However, ambiguity in expression is not necessarily an undesirable quality. Because the gestures are participants in the expressions, rather than stand-ins, the ambiguities reflect new ways of conceiving of mathematics that are emerging for the student, in the same way that ambiguities of meaning in art reflect new ways of seeing.

An example of convergence in this episode occurs in line 14, which is the second time that Crystal refers to the graphical representation. Crystal says "once that bump go forward" as she points forward with her index finger and draws a bump in the air (see Figure 10.28).

14

once <u>that</u>
<u>bump</u> go for-
ward

Crystal
maintains her
pointing
gesture while
tracing a
bump in the
air with her
forefinger.

Figure 10.28.   (00:25.0).

The phrase "that bump go forward" coupled with pointing and tracing gestures, form an utterance that is ambiguous. Does Crystal mean to show the bump moving forward like a wave or the car moving forward? Certainly, this might be unclear to an outside observer. We interpret Crystal's utterance as indicating that she was perceiving an intimate connection between bumps on the graph and forward boosts in the motion of the car. Thus, the simultaneity may indicate that a new way of seeing and thinking about force and acceleration was emerging within Crystal's lived-in space.

The above ambiguity also highlights the untranslatable nature of expressions. What Crystal is trying to express seemed fairly clear to the researchers who were present at the moment. However, to separate Crystal's utterance into the part that represents the graph and the part that represents the motion of the car is difficult, just as it may be difficult to map the various parts of a cubist portrait to the parts of an actual person.

Transitional Smoothness. In this episode as well we note a remarkable smoothness of transitions between the various representations at play. Acceleration was represented by a push or a pull on the force sensor, by the motion of the car along the track, as well as in graphical form. The smoothness with which Crystal's utterances between these representations suggests that she was beginning to recognize family resemblances between them. Since no single representation captures every aspect of a mathematical idea, multiple representations provide students with a variety of experiences and, hence, the potential to develop more complete understandings. Family resemblances are important features that connect the multiple representations of a mathematical idea. Smoothness in transitions also speaks to the role of expression in students' utterances. Crystal's sequence of expressions is joined by family resemblances across them. For example, in turn 6 Crystal says, "the other bump comes up." As she says this, she traces a small bump and then a big bump in the air with her forefinger.

6

the other
bump comes
up

Crystal traces
a big bump in
the air with
her forefinger.

Figure 10.29.   (00:11.4).

Immediately following the utterance in turn 6, Crystal makes a smooth transition to referring to the car. She says, "I think it's gonna go slo:::owly." At the same time, she flattens her hand, orienting it with the palm down, and she moves it slowly and steadily forward.

7

I think it's
gonna go
slo:::owly

Crystal's hand
assumes a
claw shape
and then
flattens,
shifting to a
palm-down
orientation.
Her hand
slides forward
as she says
"slo:::owly."

Figure 10.30.   (00:13.3).

Crystal is not looking at the car or track now but at her hand. It seems clear that she intends her hand now to represent the car and its motion. Thus, force/acceleration now seems to have transitioned from being represented by a graph (via Crystal's tracing gesture) to being represented by the motion of the car (via the motion of Crystal's hand). This is only one of several smooth transitions that occurred in this episode alone.

Interpersonal Transactions. We also see evidence that Crystal's lived-in space includes a layer of interpersonal transactions that facilitate communication about representations. A student's actions, intentions, and interactions with others, are an important component of her lived-in space. For example, in turn 1, Crystal says, "I think that it's gonna" and makes a rocking motion.

1

I think that
it's gonna

Crystal rocks in her chair, leaning nearer to the track as if to position herself to look along the track.

Figure 10.31.   (00:00.8).

Crystal's body rocking coincides with a delayed utterance. This may serve several purposes. When a speaker searches for a word, gestures help "prime a word's firing potential more quickly" (Butterworth & Hadar, 1989, p. 97) or facilitate searching for the word by providing alternate pathways to their lexicon. In rocking herself to a position where she can look down the track, Crystal is moving the track into her peripersonal space (Làdavas, Di Pellegrino, Farnè, & Zeloni, 1998), which may facilitate imagining the motion of the car because it brings the scene into her immediate view. Butterworth and Hadar (1989) also suggest that such suppressed utterances and simultaneous gestures act as interruption suppression signals that cue the other actors that the speaker is maintaining her turn to talk, while she prepares to do so. Thus, Crystal's body motion seems to play a role in the social orchestration of her interactions with the other actors.

The adoption of gestures that are specific to this mathematical environment and to the tools used by the community is also a significant feature of the interpersonal transactions. Consider the flat-handed gesture that Crystal uses in turn 7 and Charles uses in turn 9.

9

Now you're
<u>moving your</u>
<u>hand this way</u>

Charles imitates Crystal's gesture, moving a flat hand, palm-down, in the direction of the far end of the track. He looks toward that end of the track.

Figure 10.32.   (00:19.6).

Crystal uses the flat-handed gesture twice in her 30-second exchange with Charles. He uses it three times. It is not clear from this brief episode how representing the car and its motion with a flat hand emerged. What is clear, however, is that in this community, its meaning became accepted and was used freely. It is also clear that the hand is not only an embodiment of motion but also of Crystal's understanding of the information in the graph, since her motion prediction is based on a given graph and includes references to "bumps." Interestingly, the embodiment of the car's motion also includes a signal to the other actors as to when the speaker's turn ends. Crystal does this by momentarily freezing and then abruptly bringing her hand to rest on the table. Charles signals the end of his turn by freezing his hand position until Crystal starts to talk. Both signals mediate the interpersonal activity occurring in Crystal's lived-in space.

TOWARD A NEW METAPHOR OF LEARNING: THE STUDENT AS ARTIST

In the episodes that we present here, we highlight aspects of Maria and Crystal's activity that seem to us better described in terms of expression, rather than representation. We present examples from both episodes of participation, convergence, transitional smoothness, and interpersonal transactions as characteristic features of students' embodied activity in an environment of multiple representations. We now draw explicit parallels between these features of students' activity and the context of visual art. With artistic expression as our lens on students' activity, we arrive at a perspective of the student as artist. In this view, the student communicates aspects of her lived-in space via expressive acts. Taken together, a collection of these expressive acts forms a whole: the painting that the student-as-artist creates. By observing and interacting with students over time, as we did in the teaching experiment, we see these paintings develop in a layered fashion. They may have an initial composition, but that composition can transform significantly over time.

In this metaphor of student-as-artist, each of the features that we identified in the episodes manifests itself as a component of artistic expression. Participation is analogous to the evidence of the creative process that is captured in the painting. For example, in the works of Jackson Pollock, one can see traces of his action of painting. The painting serves as a record of the artist's experience in the moment. In the same way that the quality of brush strokes or paint splatters are expressive, the way that Maria's balancing gestures were enacted expressed more than just the notion of the necessity of a balance to achieve a stop. The gesture conveyed a sense of *how* she understood that balance, of what her experiences of balance and balancing had been like. In this way, the collection of Maria's bodily expressions preserved on video paints a picture of her lived-in space. In this lived-in space, balance is not merely a matter of specifying conditions that need to be satisfied in order to achieve a stop; it is an active physical process, and the nature of the act of balancing is a significant part of her understanding of balance.

With the lens of student-as-artist, we see convergence in the sense that representative elements may be brought together more closely in students' lived-in spaces than in the represented world. When representative elements are blended together in a painting, this suggests to an observer a sense of similarity between elements that might normally be regarded as quite distinct. In the episodes, we see examples of fusion, wherein two representations are brought together in a single gesture. While such expressions may be somewhat ambiguous to an outside observer, these examples of fusion suggest to us that the different representations converged in the students' lived-in spaces. In other words, they became more similar—more closely connected—than they might normally be considered.

In the context of painting, we can see transitional smoothness within a continuous brush stroke. For example, in areas within the Basquiat painting (Figure 10.1), the artist's brush strokes appear simply as brush strokes; we see them as the medium itself. They may have expressive qualities but not appear representative of any thing. In other areas, we can identify recognizable images. When looking at these, we may tend to forget the medium and focus on that which is represented. The transitions between these places may happen fluidly, from representation to medium to representation again, or from representing one thing to representing another. One cannot identify the exact point at which such transformations occur. In the experiment, we saw this kind of transitional smoothness in the role played by Crystal's hand as it fluidly transitioned from being just her hand, to tracing out a bump, to representing the car, to embodying the car's motion. In the same way that an artist can apply her brush to a canvas and smoothly transition from representing one thing to another or nothing at all, we see this same kind of transitional smoothness in students' bodily expressions.

There are also interpersonal transactions that take place during the process of creation, as well when a completed work of art is viewed. These include interactions, influences, motivations, inspirations, reactions, and so on. The language of communication is situated in the greater context of the culture and is shaped by shared norms. Interpersonal transactions encompass interactions the artist has with others during the creation of a painting, as well as after the fact, via the painting. If these aspects of an artist's experience are not well known to us, it is only because they seldom get attention.

In the metaphor of student-as-artist, everyone who interacts with the student may be a potential influence, collaborator, or audience member. For example, in Crystal's interaction with Charles, we see how her gesture of the car's motion evolved to communicate direction. This directional element was introduced by Charles as a modification of Crystal's initial gesture. She then modified the gesture again, incorporating finger pointing to indicate direction. In this way, Charles served as a collaborator with Crystal as she created the bodily expressions that communicated information about a particular corner of her lived-in space. Although Crystal modified her gesture in order to facilitate communication with Charles, the final expression was nonetheless her own.

We find it compelling to consider the pedagogical implications of the perspective of student-as-artist. In contrast to, say, the view of student-as-novice, this perspective suggests a recognition of and respect for all that the student brings to the creative process of learning. Furthermore, this perspective speaks to the role of assessment. If a collection of the student's expressions is regarded as a painting, how might her understanding be assessed? Our perspective is informed by recognition of the expressive qualities of a work of art. Unless the painting to be assessed is an attempt to mimic a particular masterpiece, it would seem inappropriate to use a masterpiece as a measuring stick. Rather, the painting ought to be assessed on the basis of its particulars, which together constitute its meaning. Thus, to assess a student's mathematical understanding would entail both immersing oneself in the details and standing back and take in the big picture, as one would in regarding a painting.

Note that we are not suggesting that some kind of radical subjectivity be brought to the difficult task of assessment. Rather, we wish to point out that the constrained nature of conventional forms of assessment may drastically inhibit the means by which a student can express her understanding of a particular topic. For example, how could Maria's understanding of balance be conveyed by her answer to a multiple-choice or fill-in-the-blank question regarding the area under the curve for a given acceleration-time graph? Her bodily experience in an environment of multiple representations afforded her the opportunity to develop a rich and nuanced sense of the relationship between force applied to the sensor, the movement of the car, and the shape of the graph of acceleration/force over time. Suppose Maria answers correctly. How is her understanding of acceleration to be distinguished from another student who received instruction aimed directly at answering test questions of particular forms? Much of the value of learning of the kind that Maria and Crystal experienced in the teaching experiment is obscured under the constraints of narrow means of assessment. Without allowing students the opportunity to express the nature of their lived-in spaces, how can we claim to assess their understanding? Too often, student learning is assessed by means analogous to asking a young artist to demonstrate her ability to color inside the lines. These implications extend to research that seeks to compare technologically enhanced instruction or kinesthetic activities with traditional instruction or other alternatives. The means of assessment may block from view exactly those aspects of students' experiences that would otherwise reveal profound differences.

Looking back at the history of Western art, accuracy (or sometimes deliberate inaccuracy) of representation was a high priority. Take microscopic realism, as exemplified by Jan Van Eyck's *Arnolfini Portrait*, 1434. The primacy given to accuracy continued into the nineteenth century. Then the camera was invented. A photograph can capture an observable scene in the real world more accurately than even Van Eyck. What then was left of the artist's role? It seems that the advent of the camera led artists to rethink the purpose of art, freeing them up to explore new possibilities. So, we see impressionism in the late nineteenth century. Artists recognized subjectivity and tried to capture how things looked to them at the moment, rather than in some objective, absolute sense. Around the turn of the century, artists experimented with cubism—an attempt to capture what the camera could not. In the early to mid-twentieth century, we saw abstract expressionism, which threw out representation altogether.

The advent of new technology resulted in a profound shift in the role of representation in art. It changed the way that artists painted, and this necessitated a change in the way that the audience looked at visual art. We see a connection between the effect of new technology on the art world and the phenomena that we observed in our teaching experiment. We are not looking at episodes of traditional instruction involving representations. We are looking at students engaged in modern, technologically influenced learning activities, which did not include direct instruction. Students behave differently in such an environment as a natural result of the new possibilities afforded by the technology. This necessitates new ways of understanding their behavior. In both art and education, the role of expression helps us make sense of new developments that challenge traditional notions of representation.

NOTES

1. Notational conventions in the transcripts:
    :: sound stretch or vowel elongation
    -      cut off word
    (.)   micro-pause
    (--)   pause, each dash represents one-tenth of a second
    [two events occur simultaneously
    (())   transcriber's descriptions
    Underlined words highlight that they were pronounced while the gesture described was being performed.

REFERENCES

Hiebert, J., & Carpenter, T. P. (1992). Learning and teaching with understanding. In D. A. Grouws (Ed.), *Handbook of research on mathematics teaching and learning* (pp. 65–100). New York: Macmillan.

Kaput, J. (1987). Representation systems and mathematics. In C. Janvier (Ed.), *Problems of representation in the teaching and learning of mathematics* (pp. 19–26). Hillsdale, NJ: Lawrence Erlbaum Associates.

Kemp, G. (2003). The Croce-Collingwood theory as theory. *Journal of Aesthetics and Art Criticism, 61*, 149–157.

Làdavas, E., Di Pellegrino, G., Farnè, A., & Zeloni, G. (1998). Neuropsychological evidence of an integrated visuotactile representation of peripersonal space in humans. *Journal of Cognitive Neuroscience, 10*, 581–689.

Merleau-Ponty, M. (1989). *Phenomenology of perception*. London: Routledge.

Meira, L. (1998). Making sense of instructional devices: The emergence of transparency in mathematical activity. *Journal for Research in Mathematics Education, 29*, 121–142.

Myers, N. (2008). Molecular embodiments and the body-work of modeling in protein crystallography. *Social Studies of Science, 38*, 163–199.

Nemirovsky, R., Tierney, C., & Wright, T. (1998). Body motion and graphing. *Cognition and Instruction, 16,* 119–172.

Noble, T., Nemirovsky, R., Wright, T., & Tierney, C. (2001). Experiencing change: The mathematics of change in multiple environments. *Journal of Research in Mathematics Education, 32,* 85–108.

CHAPTER 11

GESTURE, INSCRIPTIONS, AND ABSTRACTION

The Embodied Nature of Mathematics or Why Mathematics Education Shouldn't Leave the Math Untouched

Rafael Núñez

An essential question in mathematics education is how to improve the teaching and learning of mathematics. A tremendous amount of efforts and resources are dedicated to provide answers to this question, from curriculum planning and teacher development, to textbook design, software development, evaluation methods, and classroom dynamics. In the quest for providing answers, mathematics education, unlike other domains of teaching and learning, usually proceeds leaving the very subject matter—mathematics—untouched. Whereas other domains in education such as music, language, and literature, implicitly and naturally see the human nature of the subject matter involved in the teaching, mathematics education does not. An important factor is the widely spread view in our culture that mathematics is a transcendentally objective body of knowledge,

Mathematical Representation at the Interface of Body and Culture, pp. 309–328
Copyright © 2009 by Information Age Publishing

which exists independently of human beings, or, at best, that it is the "only truly universal language."[1] This view is endorsed—unquestioned— by many in the academic world as well as in pop culture (e.g., film, adver- tising, and video game industries). Thus, mathematical facts, theorems, definitions, proofs, notations, and so on, are largely taken as pre-given disembodied facts, external to human beings (e.g., Núñez, 2005). Conse- quently, most school mathematics is taught, generation after generation, in a relatively dogmatic form, where the very mathematical facts are rarely (or never) questioned. And I am not exaggerating. Simply think of the simple "rule" we all learn at school that "negative times negative yields positive." We all become more or less pretty good at using the rule. But are we able to explain what is the *meaning* of such statement? Or, *why* is this rule true? Or, *what* makes it true? What do you, as a reader interested in education, have to say about this question? *Why* is the above statement true, and what does it *mean*?

The fact is that most educated people I have talked to, have no clear idea of what such rule really means, or why is this rule "true." Moreover, most people I have talked to say that teachers never told them the why's. And it appears that the educational system has teachers teaching this "truth" by simply following curriculum-driven textbooks that do not men- tion the underlying meaningful *why*'s. This is hardly an isolated example. It is the rule rather than the exception. So, why is it that mathematics education does not get *at* the mathematics *itself*, and teaches its contents in a way that is friendly to known forms of human sense making?

In this chapter I argue that contemporary mathematics education would take an important step forward if, in the process of searching how to improve the teaching and learning of mathematics, it addressed *mean- ing* in a more fundamental way, and if it actually got into the mathematics itself (see also Davis' chapter 9). I suggest that one way of doing so is by informing its goals and procedures with relatively recent developments in the cognitive science *of* mathematics, that is, the scientific study of what mathematics ideas and facts are and what mechanisms of human imagina- tion make them possible. Taking a case study from foundational ideas in calculus, such as continuity of functions, I focus on embodied cognitive mechanisms such as image schemas, conceptual metaphor, notation sys- tems, and co-speech gesture production.

WHAT MAKES THE TEACHING OF CONTINUITY DIFFICULT?

To illustrate my arguments let us take a look at a specific mathematical content that in mathematics education is known for being elusive, difficult for teachers to teach and for students to learn: limits and continuity of functions (e.g., Freudenthal, 1973; Núñez, Edwards, & Matos, 1996). Are these concepts intrinsically difficult to learn? And if yes, why are they difficult? Is it that the methods used to teach them are not appropriate? Or is there something cognitively unfriendly about the very mathematical ideas themselves that needs to be understood?

Let us start by asking what is continuity. The usual definition of continuity we find in textbooks reads as follows:

- A function f is continuous at a number a if the following three conditions are satisfied:

  1. f is defined on an open interval containing a,
  2. $\lim_{\to a} f(x)$ exists, and
  3. $\lim_{x \to a} f(x) = f(a)$.

Where by $\lim_{x \to a} f(x)$ what is meant is the following:

> Let a function f be defined on an open interval containing a, except possibly at a itself, and let L be a real number. The statement
>
> $$\lim_{x \to a} f(x) = L$$
>
> means that $\forall \, \varepsilon > 0, \exists \, \delta > 0,$
>
> such that if $0 < |x - a| < \delta,$
>
> then $|f(x) - L| < \varepsilon.$

We can see that formal mathematics defines continuity in terms of limits, and limits in terms of expressions that use universal and existential quantifiers over real numbers (e.g., $\forall \, \varepsilon > 0, \exists \, \delta > 0$), and the satisfaction of certain conditions which are described in terms of arithmetic difference (e.g., $|f(x) - L|$) and smaller than relations (e.g., $0 < |x - a| < \delta$). The statements are clear and unambiguous. But, why do they create difficulties in students' minds?

An important part of the mathematics education community attributes the difficulties of teaching and learning the concept of continuity to problems related to the use of existential and universal quantifiers. Nowhere is this position clearer than in the work of the famous mathematics educator Hans Freudenthal (1973), who writes:

> The difficulties implicit in the continuity concept are quantifiers and the order of quantifiers of different kinds.... Continuity of f means intuitively: small changes of x correspond with small changes of $f(x)$. Or: if x changes little, $f(x)$, also changes little. Words like "small," "big," "little," "much," "short," "long," may hide a quantifier, but formal linguistic criteria are often

insufficient to know which kind. Always, sometimes, everywhere, somewhere—exhibit clearly the universal or existential quantifier, but the linguistic formulation does not unveil that in the continuity definition the second small (or little) hides a universal, and the first an existential, quantifier. To grasp it, a logical analysis is badly needed. The meaning of the quantifiers in "small" or "little" is better indicated in the more exact formulation: to sufficiently small changes of x correspond arbitrarily small ones of $f(x)$. Or: if x changes sufficiently little, $f(x)$ changes arbitrarily little. Still from this formulation it is a long step to understand that first the "arbitrarily little" must be prescribed before the "sufficiently little" is to be determined.... The intuitive continuity definition involves two difficulties of formalizing—first, decoding of hidden quantifiers, second, settling the order of quantifiers of different kinds. Good didactics should at least separate these difficulties from each other. (p. 561)

In the pages following this citation, Freudenthal provides an insightful analysis of these difficulties and gives helpful recommendations for achieving good didactics regarding the use of quantifiers. Whereas his analysis is precise, deep, and clear, it only focuses on the formal aspects of the ε-δ definition, missing the fundamental dimensions of everyday human cognition that may be interfering with the conceptualization and understanding of the very formalization. Freudenthal (like many in mathematics education) takes the ε-δ definition for granted, thus perpetuating the common belief that you can teach mathematics while leaving it somehow untouched or unquestioned. His analysis *starts* at the level of the formalization. This, to the point that for him "Continuity of f means intuitively: small changes of x correspond with small changes of $f(x)$." The statement is indeed a very good linguistic formulation of what is intuitive in the ε-δ *definition*. It has the right concepts relative to *that* definition, and it provides the right relations between them. What is missing, however, is precisely the need of questioning the very mathematics itself. That is, questioning the very mathematization of the everyday notion of continuity—*natural continuity* (Núñez & Lakoff, 1998), which underlie the technical notion expressed in ε-δ terms. As we shall see, a detailed cognitive analysis of the semantic organization of the ideas of *natural continuity* (present in everyday cognition) and ε-δ *continuity* (the technical definition stated above) reveals that, beyond the problem with quantifiers, these two ideas are in fact *different*, with different semantic structure and inferential organization. Not only that, these two ideas are in fact cognitive opposites: natural continuity is intrinsically holistic and dynamic, while ε-δ continuity is atomistic and static. This suggests that learning, and getting good at mastering the ε-δ technical characterization of continuity requires a considerable extra effort that goes beyond the use of quantifiers, which mathematics education should be able to address and account for. But for this, mathematics education must get at the very mathematics itself, and not take the mathematics and its formalizations for granted.

MOTION, STASIS, AND FORMALIZATION

A close inspection of mathematics textbooks reveals that usually right before giving the above formal ε-δ definition of continuity, a paragraph or two are dedicated to the so-called "informal" characterization of the idea of continuity, one that appeals to an "intuitive" description. For instance, the classic Russian book *Matematika, ee soderzhanie metody i znachenie* [Mathematics, its Contents, Methods and Meaning] by A. Aleksandrov, A. N., Kolmogorov, and M. A. Lavrent'ev (1956/1999) says: "The general idea of a continuous function may be obtained from the fact that its graph is *continuous*: that is, its curve may be drawn without lifting the pencil from the paper" (p. 88, emphasis added). And a standard calculus textbook, while discussing the same topic, says: "In everyday speech a 'continuous' process is one that *proceeds without gaps* or interruptions or sudden changes. Roughly speaking, a function $y = f(x)$ is continuous if it displays similar behavior" (Simmons, 1985, p. 58, emphasis added).

In both texts, we observe a characterization of continuous functions given in *dynamic* terms. In both cases there is something *moving*: the pencil drawing a curve on the paper in the former, and something unfolding without gaps in the latter. In both cases we have something moving from some position in space towards some other location in an uninterrupted manner. In both books these dynamic descriptions are given as a way of helping the reader by providing some immediate intuitive idea of what a continuous function *means*. The Russian book even characterizes the meaning of a "continuous function" *in terms of* something that *is* "continuous," whose meaning corresponds to what Simmons' Calculus textbook characterize as "everyday speech." At this point, right after setting this introductory presentation of continuous functions in dynamic terms, textbooks usually make a radical move. In a somewhat downgrading tone they make clear that the "intuitive" examples given so far are "merely illustrative," that they are not precise enough, and that a rigorous formal definition is required. Simmons' (1985) textbook, for instance, says: "Up to this stage our remarks about continuity have been rather loose and intuitive, and intended more to *explain* than to *define*" (p. 58, emphasis added).

This is a remarkable and profoundly informative passage. The choice of the words "explain" and "define" is not random. It characterizes the widespread idea in mathematics education that "explaining" may be a good thing, but what mathematics teaching really is about, is in "defining" entities and properties in a rigorous and precise way (in this case, presumably through the use of existential and universal quantifiers over real numbers and by establishing precise inequalities). From the perspective of cognitive science, teaching focusing on defining rather than on explaining goes against most of what is known about how humans learn and make sense of things, from perception, attention, and memory, to categorization and problem solving.

Natural Continuity, the Source-Path-Goal Schema, and ε-δ Continuity

From textbooks such as those cited above, we can see that mathematics education makes the following claims:

1. the ε-δ definition of continuity makes the so-called "informal" conception of continuity (i.e., natural continuity) *rigorous* and *precise*.

2. the ε-δ definition of continuity *generalizes* the so-called "informal" conception of continuity (i.e., natural continuity).

Let us analyze these claims separately. First, is natural continuity rigorous and precise? The so-called "informal" conception of continuity (natural continuity) is the one conceived by the creators of calculus Leibniz and Newton in the seventeenth century, and in fact, all mathematicians up to the nineteenth century (and, no surprise here, it is this notion of continuity that is brought forth over and over by students and teachers into the classroom today). It is natural continuity that brought Euler to refer to a continuous curve as "a curve described by freely leading the hand" (cited in Stewart, 1995, p. 237), and Kepler to measure "an area swept out by the motion of a (celestial) point on a physical 'continuous curve' " (Kramer, 1970, p. 528). Natural continuity—continuity as we normally conceive it outside of mathematics—is based on a *source-path-goal schema*, a fundamental cognitive schema concerned with simple motion along trajectories which has the following elements (Lakoff & Núñez, 2000): (a) a trajector that moves; (b) a source location (the starting point); (c) a goal—that is, an intended destination of the trajectory; (d) a route from the source to the goal; (e) the actual trajectory of motion; (f) the position of the trajector at a given time; (g) the direction of the trajector at that time; and (h) the actual final location of the trajector, which may or may not be the intended destination.

The source-path-goal schema is quite generic in nature and can be extended in many ways: the speed of motion, the trail left by the thing moving, the scale of motion, obstacles to motion, forces that move one along a trajectory, additional trajectors, and so on. The schema is topological in the sense that a path can be expanded, shrunk, or deformed and still remain a path and it has an internal spatial logic and built-in inferences (see Figure 11.1). For instance, if a trajector has traversed a path to a current location, it has been at all previous locations on that path; if the trajector moves from *A* to *B* and from *B* to *C*, then it has traveled from *A* to *C*; if there is a direct route from *A* to *B* and the trajector is moving along that path toward *B*, then it will keep getting closer to *B*; if *X* and *Y* are traveling along a direct path from *A* to *B* and *X* passes *Y*, then *X* is further from *A* and closer to *B* than *Y* is; and so on. As we can see, the ensemble of entailments involving the source-path-goal schema is very precise (i.e., the entailments themselves are not ambiguous).

Natural continuity builds on the source-path-goal schema, and as a result has the following essential features in its inferential organization (Núñez & Lakoff, 1998): (a) continuity, traced by motion, takes place over time; and (b) the trace of the motion is a static holistic line with no "jumps." If we take all these properties together, we can see that in terms of inferential organization, natural continuity is quite precise. The list of entailments is rich, specific, and unambiguous. In fact the level of

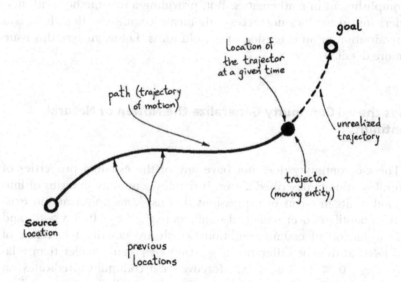

Figure 11.1.   The Source-Path-Goal schema. We conceptualize linear motion using a conceptual schema in which there is a moving entity (called a trajector), a source of motion, a trajectory of motion (called a path), and a goal with an unrealized trajectory approaching that goal. There is a logic inherent in the structure of the schema. For example, if the trajector is at a given location on a path, then the trajector has been at all previous locations on that path (Lakoff & Núñez, 2000).

precision is such that it constituted an essential building block for the invention of calculus in the seventeenth century. How about rigor? In terms of rigor, natural continuity is characterizable in terms of the source-path-goal schema with all required rigor, and it is able to handle all mathematical objects existing up to the nineteenth century (for details, see Núñez & Lakoff, 1998). Now, it is true that around the mid-1800s new, more elaborated mathematical objects emerged (e.g., "pathological" functions) that pushed for new methods to handle them. As a result the work of Cauchy, Weierstrass, Dedekind and others brought forth the ε-δ method with the specific goal of banning motion (thought to be intuitive but misleading and fallible) from the foundation of analysis. But, the point is that these were *new methods* (and very effective ones!) especially concocted for handling the cases natural continuity already handled *plus* the new untractable cases. These new methods recruited different cognitive mechanisms for its realization, such as the ideas of *preservation of closeness* and *gaplessness*. The ε-δ definition thus did not add rigor and precision to natural continuity, but rather it created a new rigorous and precise method for handling a larger universe of functions and mathematical entities. This, of course, is a great accomplishment in mathematics. But, providing a new method with new underlying ideas does not necessarily mean, cognitively, that there is a generalization or an extension of the old ideas. Let us analyze this issue in more detail.

Does the ε-δ Continuity Generalize the Notion of Natural Continuity?

The ε-δ continuity does not have any of the essential properties of natural continuity described above. It defines continuity in terms of limits, and limits in terms of expressions that use *static* universal and existential quantifiers over *static* real numbers (e.g., $\forall\, \varepsilon > 0,\ \exists\, \delta > 0$), and the satisfaction of certain conditions which are described in terms of *motion-less* arithmetic difference (e.g., $|f(x) - L|$) and smaller than relations (e.g., $0 < |x - a| < \delta$). Moreover, ε-δ continuity predicates on atomic entities (i.e., discrete points) rather than on holistic ones, such as lines. The nature of these two ideas, natural continuity and ε-δ continuity, is just radically different. In fact, with a relatively simple example, we can see that the ε-δ definition of continuity *does not* characterize the intuitive meaning of natural continuity, and in this sense it does not generalize natural continuity.

Consider the function $f(x) = x \sin 1/x$ whose graph is depicted in Figure 11.2.

$$f(x) = \begin{cases} x \sin 1/x & \text{for } x \neq 0 \\ 0 & \text{for } x = 0 \end{cases}$$

According to the ε-δ definition of continuity this function is continuous at every point. Indeed, for all x, it is always possible to find the specified ε's and δ's to satisfy the conditions for preservation of closeness (i.e., conditions expressed in static statements that indicate position within a range of closeness). However, according to *natural continuity* this function is *not* continuous. The inferential organization of natural continuity requires that certain conditions have to be met. For instance, in the semantics of a naturally continuous line we should be able to tell how long the line is between two points. We should also be able to describe essential properties of the motion of a point along that line. With this function it is not possible to do that. Since the function "oscillates" infinitely many times as it "approaches" the point (0, 0) we cannot really tell how long the line is between two points located on the left and right sides of the plane. Moreover, as the function approaches the origin (0, 0) we cannot tell whether it will cross from the right plane to the left plane "going down" or "going

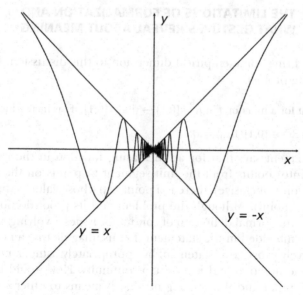

Figure 11.2. The graph of the function $f(x) = x \sin 1/x$.

up." As a result, the function violates two essential properties of natural continuity and therefore it is not continuous.

This leads us to think that the reported difficulties students have in learning limits and continuity then may not be so much due to a lack of mastery in the use of universal and existential quantifiers as Freudenthal (1973) has suggested, but to the fact that the formal ε-δ definition of continuity (a) simply does not capture the inferential organization of the human everyday notion of continuity (natural continuity), and (b) contrary to what is claimed in most mathematics books and textbooks, it *does not generalize* the notion of continuity either. The function $f(x) = x \sin 1/x$ is ε-δ continuous but it is not naturally continuous.

The moral here is that what is characterized formally in mathematics leaves out a huge amount of precise inferential organization of the human ideas that constitute mathematics. In the next section we will see that this is indeed what happens with the dynamic aspects of natural continuity that entail precise concepts such as "approaching," "tending to," "oscillating," and so on. Motion, in these cases, is a genuine and constitutive manifestation of the nature of these mathematical ideas. And, as we said, it played a key role in the work of Leibniz, Newton, Euler, Kepler, and many others. With respect to pure mathematics, however, the essential dynamic components of the inferential organization of these ideas are not captured by the ε-δ formalisms and the axiomatic system for real numbers.

THE LIMITATIONS OF FORMALIZATION AND WHAT GESTURES REVEAL ABOUT MEANING

Let us now bring some empirical dimension to this discussion. Consider the following problem[2]:

> Prove that for a function f in R^2 $f:[0,1] \to$ (into) $[0,1]$, if f is increasing,
>
> it implies $\exists x_0 \in [0,1]$, $f(x_0) = x_0$.

The statement says that for an increasing function in the real plane, mapping (into) points from the unit segment to points on the unit segment (endpoints inclusive), there is a point (x_0) whose value is that of $f(x_0)$ (called fixed point). Although the problem and its proof do not directly make use of the formal definition of continuity, issues involving continuity are deeply embedded in the statement. Let us imagine that a mathematician effectively proves this statement by appropriately using, among other concepts, the notation and idea of ε-δ continuity. How would he or she explain the proof and the *meaning* of what it means to other colleagues, say, in mathematics education? If the language and concepts of the ε-δ

formal method fully expresses the meaning of the above statement, then our mathematician will most likely unfold the explanation exclusively in terms of static universal and existential quantifiers over motion-less numbers and inequalities of numerical expressions. Consequently, one would expect the explanation not to contain elements form natural continuity such as source-path-goal motion, trajectors, trajectories, entities "approaching" others, or "tending" to other locations, and so on, elements which the creators of the ε-δ method specifically wanted out of the realm of analysis.

The question of what our mathematician would do in providing the explanation is an *empirical* one, not a mathematical or philosophical question. It is of particular interest to analyze such explanation *in vivo* and in real time. That is, rather than explicitly asking him/her "how would you explain this statement and its proof," or "how do you think we should explain the meaning of this statement to students"—which require introspection, meta-analysis and reflection on personal beliefs and behaviors—one should simply ask him/her to just explain the statement and its meaning (to, say, interlocutors in mathematics education with experience in undergraduate education). Then, one should carefully observe and analyze the gestalt of the production of such explanation, from actual notations and drawings, wordings, and verbalizations, and, most interestingly, speech–gesture co-production.

The study of gesture production and its temporal dynamics is particularly interesting because it reveals aspects of thinking and meaning that are effortless, extremely fast, and lying beyond conscious awareness (therefore not available for introspection). Rather than getting impressions of what people think about their thinking, through embodied speech-thought-gesture co-production one gets at the actual thinking in real time. Research in a large variety of areas, from child development, to neuropsychology, to linguistics, and to anthropology, has shown the intimate link between oral and gestural production. It has been shown that the phenomenon that gestures accompany speech is universal (Núñez & Sweetser, 2006) and that gestures are less monitored than speech. Speakers are often unaware that they are gesturing at all (McNeill, 1992). Besides, finding after finding has shown that gestures are produced in astonishing synchronicity with speech, that in children they develop in close relation with speech (Bates & Dick, 2002), and that gestures are co-processed with speech. For instance, stutterers stutter in gesture too (Mayberry & Jacques, 2000). Interestingly, gestures are produced without the presence of interlocutors. Studies of people gesturing while talking on the telephone, or in monologues, and studies of conversations among congenitally blind subjects have shown that there is no need of interlocutors for people to gesture (Iverson & Goldin-Meadow, 1998), which shows how deeply these hand and body movements are engrained in the thinking process. Also, it has been shown that gestures provide complementary content to speech content in that speakers synthesize and subsequently cannot distinguish information taken from the two channels (Kendon, 2000). Finally, and crucially to this chapter, gestures are co-produced with abstract metaphorical thinking, where linguistic metaphorical mappings are paralleled systematically in gesture (e.g., Núñez, 2006). In all these studies, a careful analysis of important parameters of gestures such as hand shapes, hand and arm positions, palm orientation, type of movements, trajectories, manner, and speed, as well as a careful examination of timing, indexing properties, preservation of semantics, and the coupling with environmental features, give deep insight into human thought. Taken together, all these sources of evidence support (a) the view that speech and gesture are in reality two facets of the same cognitive linguistic reality and (b) an embodied approach for understanding language, conceptual systems, and high-level cognition.

With these tools from gesture studies and cognition, we can go back to our mathematician explaining the proof, and address the empirical question we raised at the beginning of this section: How would a mathematician explain the meaning of the above statement and its proof? The following speech-gesture-writing co-production are excerpts from a recorded session that took place at the University of California, San Diego, where the mathematician Guershon Harel provided a 12-minute explanation of such proof to fellow colleagues in mathematics education. Obviously, from such a session there are many elements that can be analyzed, but for the purposes of this chapter, we are going to focus only on a couple of passages of the video.

Harel starts by laying the main concepts underlying the problem. He begins by drawing the coordinates of the xy real plane and by specifying on the positive quadrant, the unit square. He then immediately draws— from top-right to bottom-left (and from a peripheral position, relative to the body, towards the center) the line $y = x$. He follows the drawing by saying: "so, we have a function that is increasing" (Figure 11.3).

As he says "increasing," he gestures with his right hand (palm down), with a wavy upward and diagonal movement (slightly along the line $y = x$). This gesture is co-produced with the enactment of a source-path-goal schematic notion, where the source corresponds to a generic location (x, $f(x)$) with low but positive values of x and $f(x)$, the goal of the instantiated schema corresponds to a generic location (x, $f(x)$) with greater values for both x and $f(x)$, the trajector is indexed by the external edge of the right hand, and the trajectory (path) is the trace left by the motion of the hand that is indexing the increasing values of the function as x gets increasingly greater values. The gesture is inherently dynamic and it is coupled to the

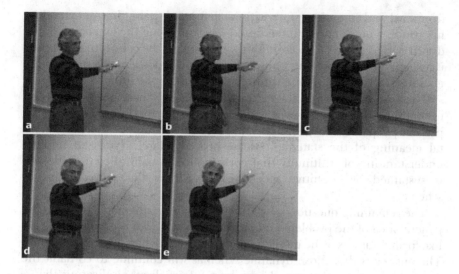

Figure 11.3.   "So we have a function that is increasing."

inscription on the board (the line $y = x$). About 45 seconds later (after discarding the trivial cases when $x = 0$ and $x = 1$, and after evoking the condition that the function is increasing) he stresses the previous idea by specifically drawing the path (or sequential paths) on the board. He says:

> But the function is increasing, so intuitively, if you think about this, it's going up (draws a slightly curved increasing line segment), it's going up (draws another curved increasing line segment following the previous one), it is going up (draws another segment, getting progressively closer to $y = x$), it is going up (draws yet another even smaller segment closer to $y = x$) ... well, at some point is going to touch here (the $y = x$ line) ... okay? ... so, something like that (and he draws another line segment on the other side of $y = x$) and then (he draws another segment) it is going to end here (corresponding to a point $(1, f(1))$ which is "below" the diagonal $y = x$) ... because it is only into. And the problem is how to capture this point [the point at which the function "touches" the line $y = x$].

The trace of the trajectory, which was drawn from the bottom left to the top right, can be seen in Figure 11.4. Although the actual tracing of the path on the board is composed of several physically disconnect segments (possibly denoting several arbitrary regions of the path), we know from the gesture production that, conceptually, the path is a holistic integrated one. Indeed, when he completes his long utterance he says "because it [the function] is only into." And he says this while producing a fast frontward and smooth gesture with his left hand (palm down), extending his arm in a quick and uninterrupted way, and along the extension of the drawn path (Figure 11.4). From this we know that the conceptualized path is a single interrupted one whose source is located where x is small to the goal when x is near 1, going through ("touching") the line $y = x$.

Then, he moves on to the next step of the explanation which is to try to "capture" the point at which the path crosses the line $y = x$. For the present purposes, what is important is that in laying out the fundamental meaning of the statement to be proved, Harel evokes an implicit understanding of continuity that corresponds to natural continuity as it is sustained by cognitive mechanisms such as the source-path-goal schema.

The remaining question is, whether once our mathematician states the general idea of the problem and moves on to work out the proof using ε-δ like formalizations, is he going to abandon these dynamic schemas or not. The answer is that those dynamic schemas will continue to co-habit the conceptualizations that are driven by ε-δ formalisms, indicating that—when continuity is implicit—there is something fundamental about the meaning of space, motion, traces, paths, and trajectories that continues to be present in the technical conception of continuity which are not grasped by the ε-δ formalisms.

For instance, after completing the formal proof of the statement, Harel stresses the crucial role played by the point at which the previously drawn path of the function "touches" the line $y = x$. (which he called x_0). When invoking the crucial least upper bound axiom of real numbers (which formally is defined in static terms only, see Núñez, 2006) with respect to x_0 he says:

> What do I mean by the least upper bound? What do I mean by this (pointing to the location $(x_0, f(x_0))$ is the edge of that set [the set A = $\{x \mid \forall x \in [0, 1] \land f(x) \geq x\}$], ... well it is the edge of that set, because if I enter the set (gestures from the right towards the left), okay? Then if I move a little bit to the left then they (points for $x < x_0$) are going to be in the set (points in the set

Figure 11.4.   Trace of the trajectory.

A) because they (the points in A) have this property [the constraints that define the set A], but if they go out of the set they are going to not to have this property anymore.

When he says "if I enter the set" he extends his right arm and at the level of the value x_0 he bends his wrist so the tip of the fingers are pointing towards him, indicating motion form the high end of the set A towards locations with lower values (see Figure 11.5), all of which have the properties specified by the set A. He closed this sentence by saying "but if they go out of the set they are going to not have this property anymore." And as he said "they go out" he points with his right index finger outwards away from his body, indicating motion towards greater values of x, values that determine points that are not members of set A (Figure 11.6). Despite the static formalization (e.g., universal quantifier and inequality signs in the definition of set A), the underlying conceptualization is carried out via the use of linear motion as characterized by the source-path-goal schema. The evidence for this comes from both, specific motor action in real time (gestures) and the properties of the co-produced linguistic expressions, which specifically recruited verbs of motion such as "to enter" and "to go out."

At this point the reader may be asking the following obvious question: Since in order to produce gestures a speaker has to move hands and arms in space, wouldn't in that case *all* gestures imply motion in one way or another? Well, the issue is that although hands and arms are moved in space they are not necessarily co-produced with a *dynamic construal*. For example, at some point in his explanation Harel uttered the expression "every set that is bounded has an edge" (Figure 11.7). At the very moment he said "bounded" he produced a bi-manual gesture (with palms towards

Figure 11.5.   "If I enter the set ..."

Figure 11.6.   "... they go out ..."

Figure 11.7.   "Every set that is bounded has an edge."

center) indicating a fixed and static one-dimensional extension where the hands indexed the bounds. Neither the gesture nor the linguistic expression uttered indicates motion (Figure 11.7a). The verbs used in this case, for instance, are the verb "to be" (existence) and the verb "to have" (possession), which do not imply motion. And a few hundred milliseconds later he completed the sentence by saying "has an edge." At the moment when he said "edge" he produced with his right hand a gesture with a hand shape resembling a horizontal letter "C," with the thumb and index finger indexing a smaller fixed and static extension (the "edge"). Again, no motion is conveyed by either the gesture or the linguistic expressions used (Figure 11.7b). So, conceptual construals can be, among others, dynamic or static. Specific properties of the linguistic expressions co-produced with specific gestures will provide evidence for one or the other.

In the case of continuity, as we saw with the previous examples, dynamic construals (linguistic and gestural) are pervasive. The omnipresent enactment of dynamic source-path-goal schemas when evoking the idea of continuity strongly suggests that dynamism is an essential part of the mathematical idea of continuity even though the so-called "generalized" concept of continuity based on the ε-δ formalisms and the static axioms for real numbers would predict that no motion should exist in such conceptualizations. What this simple case study shows is that formal languages have inherent limitations that need to be taken into account when addressing the question of the nature of mathematics and the nature of teaching and learning mathematics. The universe of human ideas is richer than the world that can be characterized with formalisms, and therefore in its efforts of improving and understanding teaching and learning, mathematics education should not take formalisms for granted, and should get at the very mathematics without leaving it untouched.

DISCUSSION

When "continuity" is taught in classrooms around the world, what is really taught? Natural continuity? ε-δ continuity? An unfortunate mixture of the two? And if ε-δ continuity is taught, is it introduced as a generalization of natural continuity? Our analysis showed that the two ideas—natural and ε-δ continuity—are different human ideas. The ε-δ continuity is also grounded on human everyday meaning, but it is based on to the static ordinary notion of *preservation of closeness near a location*: that is, being within a given distance from a specific location. And it also relies on the static ordinary notion of *gaplessness* applied to the real numbers (which is mathematized via the least upper bound axiom). Preservation of closeness has static locations, landmarks, reference-points, distances, but no trajectors, no paths, no directionalities, no motion, and therefore no "jumps." As I have argued elsewhere (e.g., Núñez & Lakoff, 1998) "preservation of closeness" and "gaplessness" are everyday human concepts with a very precise inferential organization, recruited by Cauchy and Weierstrass in the nineteenth century to carry out the program of arithmetizing analysis (Lakoff & Núñez, 2000). The cognitive science of mathematics shows that the inferential organization of the idea of preservation of closeness (plus gaplessness) is not the same as the one of natural continuity. The two concepts—natural continuity and ε-δ continuity—simply have, cognitively, two radically different logic.

As we saw above, many mathematicians and mathematics educators, thinking that mathematics education can teach mathematics leaving the very mathematics untouched, believe that the problem of natural continuity is that it is not "precise" enough. They believe that what the ε-δ definition does is to (a) make *precise* and (b) to *generalize* the idea of natural continuity. But this is not true. The idea of natural continuity is indeed very precise, and as we saw earlier, it has a precise inferential organization. The issue is that the inferential organization of natural continuity, which is dynamic and holistic in nature, did not serve the purposes of the arithmetization program which required a reduction of mathematical ideas into static and discrete numeric structures and concepts that later became compatible with static set-theoretic concepts (based on static concepts such as membership relation seen as presence inside a container, and lack of membership relations as presence outside the container, etc.). The inferential structure of the idea of preservation of closeness did fit the goals of the arithmetization program, and it is *this* idea (along with others such as *gaplessness*) that was made precise by the ε-δ definition.

The moral is that mathematics education, as an endeavor directly dealing with humans and with the question of how they think and learn, should not assume that formal definitions in mathematics (a) make intuitive ideas more "precise" and that (b) they generalize those intuitive ideas. From this it follows that contrary to what Freudenthal (and many mathematics educators) say, "the difficulties implicit in the continuity concept" are not just the quantifiers and their order. The problems are deeper. They start with the false belief (from both, the teacher and the student) that formalization necessarily generalizes and makes intuitive ideas more precise. Moreover, the problems multiply if the subject matter of the teaching—mathematics—is taken for granted, thus hiding the human nature of the discipline. Because the cognitive science *of* mathematics does indeed question the nature of the very mathematics, it is in a good position to show exactly why mathematical formalizations neither make more precise, nor generalize, everyday intuitive ideas.

Today, there are important methodological and theoretical advancements in research that help with the investigation of the cognitive science of mathematics and its embodied nature: gesture studies, cognitive semantics, eye-tracking studies, discourse analysis, ethnographic observation, neuroimaging, to mention a few. Mathematics education would benefit tremendously by building on these kinds of developments and by acknowledging, in a deep way, that mathematics is indeed the product of human imagination, and that it can be taught with *meaning* as one of its pillars.

NOTES

1. As I write these lines IBM is running a TV commercial which opens with a relatively young mathematician looking at the camera, who, while standing in front of a white board full of mathematical notations, utters unambiguously what seems to be an undisputable fact: "Math is the only language all humans being share." The clip is available at the URL http://www .youtube.com/watch?v=-udGE8POcZk&feature=channel_page

2. This problem was suggested by Guershon Harel in the context of a study on the nature of proof conducted in collaboration with Laurie Edwards. I thank both of them for stimulating conversations about this problem and the nature of proof, and for allowing me to use the video data included in this article.

REFERENCES

Aleksandrov, A., Kolmogorov, A. N., & Lavrent'ev, M. A. (1999). *Matematika, ee soderzhanie metody i znachenie* [Mathematics, its contents, methods and meaning]. Mineola, NY: Dover. (Original work published 1956)

Bates, E., & Dick, F. (2002). Language, Gesture, and the Developing Brain. *Developmental Psychobiology, 40,* 293–310.

Freudenthal, H. (1973). *Mathematics as an educational task*. Dordrecht, The Netherlands: Reidel.

Iverson, J., & Goldin-Meadow, S. (1998). Why people gesture when they speak. *Nature, 396,* 228.

Kendon, A. (2000). Language and gesture: unity or duality? In McNeill (Ed.), *Language and gesture* (pp. 47–63). Cambridge, England: Cambridge University Press.

Kramer, E. (1970). *The nature and growth of modern mathematics*. New York: Hawthorn Books.

Lakoff, G., & Núñez, R. (2000). *Where mathematics comes from: How the embodied mind brings mathematics into being*. New York: Basic Books.

McNeill, D. (1992). *Hand and mind: What gestures reveal about thought*. Chicago: Chicago University Press.

Mayberry, R., & Jacques, J. (2000). Gesture production during stuttered speech: insights into the nature of gesture-speech integration. In D. McNeill (Ed.), *Language and gesture* (pp. 199–213). Cambridge, England: Cambridge University Press.

Núñez, R. (2005). Creating mathematical infinities: The beauty of transfinite cardinals. *Journal of Pragmatics, 37,* 1717–1741.

Núñez, R. (2006). Do real numbers really move? Language, thought, and gesture: The embodied cognitive foundations of mathematics. In R. Hersh (Ed.), *18 unconventional essays on the nature of mathematics* (pp. 160–181). New York: Springer.

Núñez, R., & Lakoff, G. (1998). What did Weierstrass really define? The cognitive structure of natural and ε-δ continuity. *Mathematical Cognition, 4*, 85–101.

Núñez, R., & Sweetser, E. (2006). With the future behind them: Convergent evidence from Aymara language and gesture in the crosslinguistic comparison of spatial construals of time. *Cognitive Science, 30*, 401–450.

Núñez, R., Edwards, L., & Matos, J. F. (1999). Embodied cognition as grounding for situatedness and context in mathematics education. *Educational Studies in Mathematics, 39*, 45–65.

Stewart, I. (1995). *Concepts of modern mathematics*. New York: Dover.

Simmons, G. F. (1985). *Calculus with analytic geometry*. New York: McGraw-Hill.

EDITOR'S SECTION COMMENTARY

Wolff-Michael Roth

I begin this Part C of the book with an account of the kind of mathematics education that I received in my early years. The teaching methods focused on the reproduction of preformatted answers to textbook questions. Teachers never appeared to be interested in asking whether I, as my peers, were really engaging and identifying a problematic in the way I had done during my nocturnal efforts in mental calculations and the algebraic understanding that I developed. However, cultural-historical activity theory suggests that it is out of a concern with real problematic issue that students orient toward a motive, in the pursuit of which they will do what is required to enlarge their room to maneuver, that is, they engage in learning loops that expand their possibilities of acting. The object/motive does not and perhaps should not have to be mathematics itself, to which students will inherently turn when they can see in it something that will allow them to better work toward their chosen object/motive.

In the present set of chapters, both Rafael Núñez and Brent Davis explicitly ask the question and address the issue of the mathematics. For the former, mathematics education leaves the subject matter untouched. I see the latter in his work with teachers do precisely that when he asks

Mathematical Representation at the Interface of Body and Culture, pp. 329–331

them to conceptualize what in ordinary mathematics lessons are simple procedures, for example, multiplication. He engages them in ways that makes them co-producers of mathematics, which may subsequently lead to different classroom teaching approaches that engage the students in the production of mathematical knowledge on their part.

In the third chapter of the set authored by Ian Whitacre, Charles Hohensee, and Ricardo Nemirovsky, we can see a very different aspect of mathematical cognition that is enacted while students engage with the physical world. Mathematics educators might justly question what the mathematics in such situation has to do with the kinds of practices that we observe in pure mathematics. As one of my studies of mathematics among fish culturists showed, cultural-historical activity theory predicts very different forms of mathematical.practices in the fish hatchery than in the offices of a mathematics department. This is so not only because of the different levels of mathematical competencies. Rather, this is so because of the different orientations of the productive activities that occur in fish hatcheries and in mathematics departments (Roth, 2005). In the former, fish culturists use mathematics to model the production and growth of the annual brood, whereas mathematicians articulate mathematical objects as their motive of inquiry. Already I can see that the mathematics that Chris Rasmussen et al. describe in chapter 7, which is more like what we would expect in mathematics departments, is very different from the one that I observed in the fish hatchery, though both deal with modeling a brood of fish. Similarly, we might expect very different mathematics to emerge in the case of Whitacre et al. situation, on the one hand, and students pursuing pure mathematical objects of the same nature, on the other hand. But mathematics in this chapter 10 can come even further to the fore when it is made thematic in its role to achieve the goal or motive at hand.

In all three chapters, we see explicitly or implicitly the idea of mathematics as creative activity, where I mean creative in terms of producing something that is in excess of and not directly derivable from current knowledge. In fact, the possibility to rediscover mathematics is a condition for its appearance as a transcendentally objective knowledge that Núñez points to in his opening paragraph. In the final chapter 12 of this book I return to the issue, fundamentally taking a phenomenological perspective on it worked out by Edmund Husserl and an ethnomethodological perspective presented by Eric Livingston. Both are well situated to speak to the point, the former having done a PhD in mathematics and the latter having studied mathematicians proving Gödel's theorem. This creative aspect therefore, the possibility of discovering mathematical proofs over and over again, and its relation to the apparently transcendental and objective nature of mathematics awaits further work in mathematics education—though readers can have a taste of how to go about it in my own study of second-grade children doing geometry (Roth & Thom, 2009). I particularly like Whitacre et al.'s metaphor of students-as-artists and further work is required to better understand the tension that arises from the dialectic of production, that is, the creative reproduction and creative transformation of mathematical practices that occurs in mathematical praxis.

REFERENCES

Roth, W. -M. (2005). Mathematical inscriptions and the reflexive elaboration of understanding: An ethnography of graphing and numeracy in a fish hatchery. *Mathematical Thinking and Learning, 7*, 75–109.

Roth, W. -M., & Thom, J. (2009). The emergence of 3d geometry from children's (teacher-guided) classification tasks. *Journal of the Learning Sciences, 18*, 45–99.

PART D

EPILOGUE

CHAPTER 12

APPRECIATING THE EMBODIED SOCIAL NATURE OF MATHEMATICAL COGNITION

Wolff-Michael Roth

The stories of genius that mathematicians propagate, and the heightened sense of discovery that popularizations give, distract attention from mathematical discovery as the ordinary circumstance and aim of provers' activities. (Livingston, 1987, p. 137)

In the introductory quote to this chapter, Eric Livingston, who in his doctoral dissertation has studied a pair of mathematicians doing Gödel's proof (Livingston, 1986), notes the distraction that occurs as part of the disattention to the "ordinary circumstance and aim of provers' activities." That is, the way in which mathematics is depicted, often in the transcendental properties of its objects that are said to exist independent of the material properties of the proof accounts, we do not see the embodied work that mathematics consists of (but see Núñez' chapter 11). In fact, the study of the mathematical objects, such as a proof as it appears on paper or whiteboard, cannot ever lead us to the process that brings it about

Mathematical Representation at the Interface of Body and Culture, pp. 335–350

(Henry, 2000). This process or work is invisible in the description of mathematical cognition. That is, there are actually two parts to mathematical proof. On the one hand, there is a proof account, which exists in and through the communicative devices that make the proof available to others (e.g., in a journal). On the other hand, there is the lived work of proving, without which no proof exists. This embodied work is required not only of the person who does some proof for the first time, but also on the part of the reader, in participative understanding, who reproduces and transforms the lived work of proving. This lived work not only is embodied, it is social as well, because directed toward the (anonymous) other who, in his or her own lived work, actually reifies the independence of the proof from the material particulars in which it is presented. But a proof is a proof only if another person, using the *materials* at hand, reproduces the proof account in and through his or her own lived work, which, as such, becomes an expression of mathematical actions that another recognizes as his or her own. Mathematics is both embodied and social simultaneously, and the two aspects, the embodied work and the proof account produced, constitute a *Lebenswelt* (life-world) pair (Lynch, 1997). Mathematics, therefore, becomes transcendental precisely when its objects, the proofs, are detached from the actual details of doing proving. In contrast to the contention of some who continues to consider mathematics as a body of transcendental phenomena, the position I have been taking throughout this book to organize the chapters is one in which mathematics simultaneously is embodied and social.

This Lebenswelt pair constitutes the source of mathematics as a historical and epistemological phenomenon (Husserl, 1939). The very possibility for mathematics to be an objective science rides on the possibility to rediscover it in and through the embodied work of doing mathematics. The reigning dogma has separated epistemological and historical explanations (i.e., the origin) *in principle*. Mathematics, as a science, is not *merely* a historical tradition; its very history is grounded in the objectivity of its repeated *achievements* as cultural fact. At the same time, the continued objectivity of its cultural achievements *is* a historical phenomenon through and through. Thus,

> [j]ust as the individual cultural gestalt has behind and *in itself* its tradition, so is the entire cultural present, understood as a totality, implicated in the cultural past. More exactly, in it is implicated a continuity of mutually implicating pasts as respective cultural Umwelts. (p. 220, my translation).

Throughout this book, the contributing authors and I articulate, to varying extent, the double nature of mathematical cognition in the real world of school classrooms and interviews (think-aloud protocols). In both the embodiment and social approaches to mathematical cognition, the material nature of mathematical practices comes to the fore. In the former approach, the material nature is inherent in the approach, as consciousness itself derives from previously non-conscious articulations in and through the human body. At the heart of the social approach lies communication, inherently a material phenomenon. Using cultural-historical activity theory, I show throughout this book that there cannot be communication without a material world in and through which signs, always already material, are made materially and therefore objectively present for all participants to the situation. Because of its focus on consciousness, cultural-historical activity theory lends itself to integrate the perspectives that take the position of the actors onto the social world, that is, the phenomenological and ethnomethodological perspectives of Husserl and Livingston.

In the following, I provide an integrated and integrating perspective by first addressing some of the critiques that have been launched at the embodiment hypothesis and then elaborate the issue of how mathematics can emerge over and over again in the cultural praxis of its new adepts.

CRITIQUES OF THE EMBODIMENT HYPOTHESIS

Whereas some scholars have accepted the articulation of the embodied nature of mathematics, there also has been a response to *Where Mathematics Comes From* (Lakoff & Núñez, 2000) that was rather negative and even vicious. These "attacks" generally have come from mathematicians, who with some justification in regards to the first printings of the book, point out that there were errors in the mathematics presented. On the other hand, there are other aspects of the critique that are entirely unjustified. The source of these inappropriate readings appears to arise from a fact related to the point Rafael Núñez makes. He suggests that mathematics education leaves untouched mathematics. Mathematicians, on the other hand, argue on mathematical grounds rather than on cognitive grounds, which is the very point Núñez has made in response to his critics.

Upon first reading a number of these critiques, I was reminded of the debates in and around the sciences in response to ethnographic studies in scientific laboratories. Here, too, the scientists often claimed that the social scientists did not know what they were talking about and the scientists felt that they had a better grip on the construction of knowledge in the sciences. And they did so even after Thomas Kuhn has suggested that the history of science from the scientists' perspective is essentially revisionist, a

Whig history, explaining the course of science history in terms of its ultimate achievements. The misunderstandings, as has been shown in a study of mathematicians at work reproducing the schedule of proofs of Gödel's incompleteness theorem, arises from the differences in the objects that researchers consider. Thus, "on every occasion when the nature of mathematical activity is posed as its own topic for mathematical, philosophical, historical, or sociological reflection and analysis, the mathematical object is first disengaged from the local work to which its natural accountability and analyzability are integrally tied" (Livingston, 1986, p. 7). When the lived and embodied work is studied, which is tied to the mathematical object at hand, researchers reach very different conclusions than when they consider mathematical activity as its own topic. But let us take a look at some of these critiques. Here lies precisely the point: When the mathematical object is disengaged from actually doing mathematics—in which all chapters in this volume are interested in—then one gets to a very different, inappropriate reading of what mathematical cognition consists in.

One of the critiques, which appeared in the flagship journal *Nature*, does little more than sketch a caricature of the points of the book (Goldin, 2001). The review dwells in generalities without carefully working out in which way the approach Lakoff and Núñez offer lacks intelligibility, plausibility, and fruitfulness. The author complains that Lakoff and Núñez fail to include "essential ideas" from cognitive science, and then list the very concepts that are characterized by their idealism rather than embodiment: internal representations, information processing, cognitive structures, affect, motivation, or analogical reasons. The following quote provides a flavor of what the review is like:

> As a survey of ideas in mathematics, this book does not compare favourably with other popular expositions. Occasional misconceptions, and frequent imprecision of mathematical language in otherwise valid explanations, make close page-by-page reading frustrating. Of course, the authors are not mathematical scientists. Like students new to the subject who are striving to understand the ideas behind formal mathematics, they "discover" that multiplication by i implements a 90° rotation, that space-filling curves do not fill space (when hyper-real coordinates are included) and that symbolic logic is "not absolutely true." (p. 19)

Goldin writes with irony and glee, but does not get the real point, which is that discovery *is the condition of mathematicians' work*. The very genetic origin of the discipline, as Husserl (1939) shows, derives from the nature of mathematics as a discovery science. Moreover, the very tradition and tradability of mathematics depends on the rediscovery of mathematics across millennia. "Each proof is a discovery—the elementary ones no less than the more consequential, celebrated ones" (Livingston, 1987, p. 137). It is only because a proof can be rediscovered over and over again—always and already in and through the lived work of the investigator—that geometry can exist as a science with objects that have the appearance of transcendence.

In my reading, the review is of little use to the development of mathematical cognition, as it constitutes a superficial wholesale rejection of the embodied cognition approach without specification of exactly where the approach fails to account for the data. It is true that social science researchers who do not know a field and its forms of discourse easily misunderstand what insiders understand in saying this or that. I have experienced it while working with a social psychologist who misunderstood a physicist and a physics undergraduate student while talking about the different signage of birthrate and death rate in physics and biology. In the two disciplines different expressions are used for the combined effect of the two rates because of the ways in which these are conceptualized and understood ($b + d$, in physics, where d would be a negative quantity; $b - d$, in biology, where d is a positive quantity).

In MAA Online, Bonnie Gold provides another critique that is taking an equally high horse of the mathematician who finds errors in this or that depiction of a mathematical idea and who fails to discuss the *cognitive* implications. Her opening remarks are characteristic of a review that does little to push the cognitive science of mathematics ahead:

> *Mathematical platonists, awake!* Here, in a book written to put the final nail in the coffin of platonism, we have the beginnings of a response to the one serious philosophical challenge to platonism: how can human beings, finite physical beings, ever develop an understanding of mathematics, abstract and infinite, disjoint from physical experience? (http://www.maa.org/reviews/wheremath.html)

A much more sophisticated analysis of the metaphor approach to the embodiment of mathematical cognition is provided as part of a review a book in which Lakoff and Johnson also have a chapter (Dubinsky, 1999).

> [T]hey misinterpret standard conventions of mathematics connected with expressions such as $\lim_{x \to \infty}\left(\frac{1}{x}\right) = 0$ and $\lim_{x \to 0}\left(\frac{1}{x}\right) = \infty$, missing the point that the first asserts an equality of two numbers but the second is just a notational convention for a certain phenomenon and does not mean that ∞ is supposed to be a number. There is a long discussion of the continuity and differentiability at 0 of functions defined by expressions such as $x^n \sin\left(\frac{1}{x}\right)$.

They seem to deal with these pictorially in terms of gaps and tangents and directions. But although their descriptions work for $n = 1$, 2, they break down for larger values of n. In my view, these are not "monster" functions, as Lakoff and Núñez designate them, but rather an intriguing collection of examples that serve to ground the concepts of continuity and differentiability of different orders in ways that do not seem to be included in their metaphor epistemology. (p. 557)

The comment is revealing in many ways and does not undermine the fundamental idea of cognition as a metaphorical extension of bodily experiences in the world. First, Dubinsky notes that the first of the two expression is an "equality of two numbers" whereas the "second is just a notational convention." In fact, both are notational conventions, as mathematical operations more generally. The argument is actually circular, in that the first notation is an equality between numbers because the part to the right is a number, making the "=" a sign of equality. The second is not an equality, giving the "=" a different sense, because ∞ is not a number. Further, from a cognitive perspective, the very nature of the sign "=" is in question, and as Dubinsky points out, there is convention involved. But a convention is, as the etymology of the word makes clear (Lat. *con-*, together with + *venīre*, to come), inherently a social phenomenon, a coming *together*, an agreement.

Dubinsky exhibits uncertainty in several ways. First, he hypothesizes that Lakoff and Núñez deal with of functions pictorially ("They seem to deal with ..."), though precisely what the authors are doing has not been specified: there is no more and no less. Second, Dubinsky uses a modifier that exhibits a particular point as "his view," which means that it may not be shared at all within the mathematics (education) community. It is, as we know from the social studies of science, a hedging move that attributes probability of error to a potential factual statement. Finally, he hypothesizes about the inclusion of concepts in the "metaphor epistemology" and therefore is uncertain about this fact ("do not seem to be included"). As an alternative, Dubinsky proposes constructivism, which, as my introductory section on Kant, Piaget, and (radical) constructivism shows (see chapter 1), has some fundamental problems in its own right.

Reading Dubinsky's complaint that Lakoff and Núñez present only one way rather than the many ways in which mathematicians articulate a particular topic brought back to memory my own relation to vectors. I still remember the struggles I experienced at the junior college level (about 12th grade) while having to calculate the distance between two non-parallel three-dimensional vectors. Although I came to have a better sense for vectors while doing my MSc in physics, my difficulties did not abate, especially not in theoretical physics and especially not when we studied the matrix approach to quantum mechanics. (I had this sense even though I came to have among the highest grades in theoretical physics.) My breakthrough actually came later, when I programmed microcomputers to do polynomial regression and best-fit analyses requiring the manipulation of vectors (the data) and matrices. As soon as this new practical sense with vectors emerged, I also developed a new sense for the matrix approach to quantum mechanics and other fields where vectors and matrices where central to the mathematical models. That is, there was one experience in particular that gave me a practical understanding of the field, and by means of it, I began to connect many fields of applications and understandings that I previously did not have. The point is, that it is through the practical understanding and embodied work in one form of representation that a whole field opened up for me. Once a practical understanding has emerged, it provides a ground that leads to further success of the practical and applied aspects of the sciences, including geometry (Husserl, 1939). With this success comes an "*unavoidable change of sense*, which are the results of such a way of proceeding scientifically" (p. 217, my translation and emphasis).

There is therefore a loss of the original sense, the original *sensible* experience, the one that Lakoff and Núñez refer to as extended by means of metaphors to ground the higher order, more abstract mathematical entities. The disconnection, according to Husserl, occurs slowly, because the early mathematicians were still connected to the original sense and because of the practical value mathematics had in their lives. The problem of the *disconnect* from the sensible, embodied nature of mathematics could become evident only in the twentieth century, not only in the work of Husserl but also, and especially, in the cognitive science and artificial intelligence that face the "grounding problem." Lakoff and Núñez's work is appealing precisely because it provides an intelligible, plausible, and fruitful response to the problem that Dubinsky's and Goldin's (radical) constructivist, information-processing, cognitive psychological, or cognitive structures approaches have not been able to provide answers to: how mathematics is related to the world generally and to the person doing mathematics specifically.

One of the questions none of the critics raises is this: How did mathematics historically get off the ground from what were fundamentally nonmathematical endeavors? Similarly, how, if students bring to school nothing but their everyday practical understanding of how the world works, can they "construct" any mathematical ideas (concept) a process that requires that the mathematical idea (concept) exists for them to be an object of their intentions? Both of these questions cannot be solved by means of the Whig history that mathematicians tend to favor or by constructivism that Dubinsky asks us to rely on. To get mathematics off the ground of fundamentally nonmathematical and pre-mathematical everyday practices, we need to have a process that allows consciousness to bootstrap itself, *on the ground and by means of* pre- and non-mathematical practices and *by taking these as its object*. One such approach has been articulated nearly 100 years ago by the mathematically trained (PhD) Edmund Husserl in his search for the origins of geometry. Explaining learning from the perspective of the *subsequently* known is a fruitless and deceiving exercise, for prior to acquiring the language of mathematics, as in the acquisition of any language (Rorty, 1989), the learner cannot envision what this can be used for and what can be done with it, because the language itself provides the descriptions thereof. This is so because the new language "makes possible, for the first time, a formulation of its own purpose" (p. 13). Prior to the acquisition of the language the learner, someone who makes something truly new for himself, "is typically unable to make clear exactly what it is that he wants to do before developing the language in which he succeeds in doing it" (p. 13).

To rectify the problems in the study of mathematical cognition, we need to return the mathematical object to its origins within that culturally and historically contingent and once-occurrent lived praxis itself. In so doing we are enabled to study the natural analyzability and accountability of mathematical work of students and professionals alike in and as the accountable, inspectable and therefore analyzable, real-world practices of human beings that inherently appear in body (material) and flesh (endowed with senses for sense making).

EMERGENCE OF MATHEMATICS IN AND FROM (PRE-MATHEMATICAL) EXPERIENCE

From a historical perspective, for anything like present-day geometry to emerge, there had to be apodictic aspects in the prescientific world of the founders of the discipline, which provided the materials for idealizations that are characteristic of the discipline (Husserl, 1939). Similarly, the world of children provides them with the materials and grounds on and from which their first geometrical understandings emerges, not as something radically strange and different, but as something that makes sense from the first moment. This prescientific world, however structured, phylogenetically was and ontogenetically is invariantly, a *social* world of *material* objects (bodies) encountered with and through the human body: "The Ur-materials for the first production of sense, the Ur-premises so to speak, exist, *prior to all science*, in the world of life, which is not merely a thingy Umwelt, but already structured cultural Umwelt" (p. 219, my translation). The things are not *mere* objects, because the coexisting human beings and the coexistent cultural objects cannot be reduced to pure thinghood (Derrida, 1989). These pure bodies had, besides their spatiotemporal characteristics, also "material" qualities including color, heat, heaviness, hardness, size, and so on. For geometry to emerge, the spatial shapes, temporal shapes, and shapes of motion have to be singled out from the totality of the perceived object; their independent and concrete reality, which comprises all the material, sensible qualities and the *totality of their predicates* have to be disregarded.

Distinguishable features of objects initially are surfaces, edges, lines, corners, and points. In the praxis of working with such materials, preferential properties were continually refined and related—for example, boards that meet and are bordered by lines, points, and corners. An ideal and idealized construction—one that is valid across historical time, space, and cultures and therefore reproducible with identical intersubjective sense—can arise from all of this only under the condition that "the apodictically general content, invariant across all conceivable variations of the spatiotemporal sphere of gestalts is taken into account in the idealization" (Husserl, 1939, p. 224). This conception of the nature of geometry also is relevant and useful in the study of mathematics in a second grade (chapter 4) where geometry emerged in and through children's classification tasks (Roth & Thom, 2009b). The objective of the sorting task precisely was to arrive at a classification where objects of similar gestalts (forms, shapes) are placed together so that collections emerge with invariant spatiotemporal properties (as articulated in the predicates), differing from other collections in the spatiotemporal properties, while permitting other properties to differ and be the same within and across collections, respectively. That is, there might be red and green objects in the same class of cubes, or red cubes and red cylinders. But all objects of cubic gestalt, for example, *have to be* in the same collection, for otherwise geometry would not properly emerge.

The study from which chapter 4 (Maheux et al.) evolved began with the assumption that whatever children did could be counted correct in some classification (and therefore language) "game" and the associated rationality instead of assuming that the children were in error, to varying amounts. What is or are the games and the rationality within which their actions make sense? That is, the game and children's rationality are taken to be phenomena sui generis, not some faulty and diminutive forms of adult games and rationality. In fact, our data show that children did competently classify, just that the predicates used were not the ones permissible in the game that the teachers designed. Such an approach is consistent with the cultural-historical emergence of geometry as a field, which began with some invention of first geometry without that the first geometers could know what the total sense of geometry is at their moment in time or would be some time (years, decades, centuries) later. That is, the sense of geometry could not be a predicate of the intentions of the first geometers. Nevertheless, geometry as it exists today has evolved from the very early beginnings, which themselves have evolved from a rationality that did not include geometrical thinking. To understand the cultural-historical emergence of geometry, we cannot reason in a teleological manner—this would be Whig history—but we need to squarely address the question how anything such as geometry could have emerged from the reigning common sense that existed just prior to the first statements of geometry.

In the lessons of the second-grade children, geometry (as praxis rather than as meaningless word) begins to emerge from their tasks. From pragmatic and methodical perspectives, members to organized arrangements such as mathematics classrooms are continually "engaged in having to decide, recognize, persuade, or make evident the rational, i.e., the coherent, or consistent, or chosen, or planful, or effective, or methodical, or knowledgeable character" (Garfinkel, 1967, p. 32) of their inquiries as sorting, classifying, providing reasons, counting, graphing, sampling, reporting, and so on. What is at issue for me, therefore, is geometry-in-the-making rather than ready-made geometry as it appears in books. This is precisely the aspect that the critics of the Lakoff and Núñez work do not appear to understand. Across time, geometry, as an ideal object, can exist only in and through a second layer, the sensually embodied practices that localize and temporalize (e.g., Euclidean) geometry as non-local and non-temporal ideality. That is, the objective nature of geometry exists only in and through the emergence of geometry in the practical actions of real, embodied, sensual human beings. It is only in its production that geometry as objective science is reproduced across generations. It is not therefore that geometry lessons *cause* students to do certain things; nor is it that in their actions, students produce geometry anew. Rather, ideality and materiality of geometry as objective science emerges each time when a member produces and thereby reproduces practice that are recognizably and therefore collectively geometrical in nature. (Because idealities are taken up by new idealities, newcomers to geometry do not have to reproduce the phylogeny of geometry; a corollary is that earlier idealities are not necessarily recoverable at some later time.) In the way the sense (meaning) of language and words lies only in their use, the sense of geometry and geometrical objects solely derives from their *concrete* use for specific purposes in particular settings.

One way in which early peoples could bootstrap out of everyday into mathematical conceptions is by metaphorical extension of the concepts that they had acquired through embodied experiences. This bootstrapping occurs in the process of the pursuit of object/motives, where developed forms of mathematics expand the action possibilities. As these experiences are bound up and at the heart of each word, which is continually reproduced and transformed, each word actually gathers up human experiences over the course of millennia. Initially embodied experiences are associated with words. These words, subsequently come to metonymically denote those experiences and the situations in which they occur (Roth & Thom, 2009a). In the course of continual reproduction and transformation of language (Bakhtine/Volochinov, 1977), the words come to embody collective memory. It is therefore not surprising when many philosophers seek recourse to etymology (history of the word) to uncover some of the fundamental experiences that have entered the words that we now use without being aware of it. Thus, for example, the word cylinder derives from the Greek $Kυλινδρος$ [kulindros], meaning roller. Children learning about cylinders do, as the data from the study in chapter 4 show, roll cylinders between K their two hands when asked to describe what makes a cylinder a cylinder ("you can roll it"). The word cube comes from Greek $Kύβος$ [kubos], originally a die to play with before denoting the mathematical object we know as cube; cone comes from the Greek word $KΩνος$ [konos], pinecone; and pyramid comes from Greek $πυραμίς$ [puramis], the monumental structures that served as royal tombs in ancient Egypt. Embedded in the mathematical concepts denoted by these names there are millennia worth of human experiences, captured and preserved in the physical body of the sound and the transcriptions into the language of their use.

Each of these words has been part of (written, spoken) utterances since their first appearance and taken up over and over again in an unending "chain of speech communication, and it [word, utterance] cannot be broken off from the preceding links that determine it both from within and from without giving rise within it to unmediated responsive reactions and dialogic reverberations" (Bakhtin, 1986, p. 94). Some links to the original experiences get broken, such as in the case of the cylinder, which has no apparent link to the verb "to roll" or the noun "roller." But the experiences remain embodied in the very reproduction and transformation of the concept words.

Our experiences in and relations with the world, both material and social, therefore are the foundation of any higher cognitive function: "all the higher functions originate as actual relations between human individuals" (Vygotsky, 1978, p. 57). It is in and through societal relations that we come to know about, reproduce and transform existing thinking practices. And it is in and through social relations that we make the same practices available to others, both peers and students. In communicating, we make use of sound (words, prosody), gestures, body positions, body movements, and so on. Any form of practical consciousness, whether it relates to the natural world or mathematical objects, is consciousness-for-the-other, and therefore can be observed in the relation of people. We do not have to go into the mind and into the transcendental (i.e., not derived from experience) ether where the noumena are thought to reside.

Ethnomethodology is an ideal method for studying embodied cognition in action because of its interest

> on the ways in which accounts are essentially and irremediably tied to the activities of which those accounts are a part, and on the ways in which a local cohort's production and management of the activities in which they are engaged are identical to the ways in which those activities are made accountable. (Livingston, 1987, p. 124)

Mathematicians work with proofs. The work of proofs is inherently social work, because proofs are naturally accountable. The proofs themselves therefore are social objects, and witnessable as such.

> This is so not because some type of extraneous, non-proof-specific element like a theory of "socialization" needs to be added to a proof, but because the natural accountability of a proof is integrally tied to its production and exhibition *as* a proof. (p. 126)

It remains to be shown how the illusion of the transcendental nature of mathematics emerges, the nature that the mathematicians critical to Lakoff and Núñez take as an a priori given. The transformation of a mathematical object from a social object into a transcendental one can be seen in the following case of the visual proof of the Pythagorean theorem (Figure 12.1). Especially non-mathematicians might find it puzzling to see how the figure constitutes a proof account of the theorem. But once the figure can be seen as constituting the proof, it is transformed, instantly becoming transcendental and objective.[1] Rather than existing elsewhere,

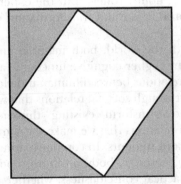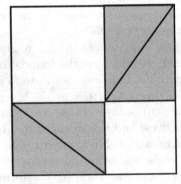

Figure 12.1.   Proof account for the Pythagorean theorem.

the knower can see the proof to exist in the figure itself and, for the first-time discoverer, the discovery itself may be a considerable aha experience. That is, the proof

> has a substantial "massive" presence. It appears to have an endless depth of proof-relevant, discoverable details; these appear to be available from different perspectival viewings of *the proof's* various aspects. It withstands its repeated interrogation and is seen to be independent of all inquiries into its properties. It is seen retrospectively as the cause and source of all inquiries concerning it. It is accountably and analyzably a proof of the Pythagorean theorem. (Livingston, 1987, p. 119)

That is, although the proof can be fully shown, thereby exhibiting the work of proofing, the figure itself comes to capture the essence of the proof as if it were independent of the embodied, lived work that the figure denotes. (This is equivalent to the case discussed in the introduction to Part A, where the perspectival perception of the cube is seen as independent of the eye movements that constitute the object in the first place.) But as Husserl (1939) suggests, we cannot understand either the epistemological or the historical aspects of a science such as mathematics when we separate history and epistemology from the content matter. It is thereby, too, that the originally sensible nature of mathematics is lost: "it is evident in advance that the new entities are products that emerge from an idealizing nature of mental doings, a 'pure thinking,' which found its material in the described givens of factual humanity and human Umwelt and *makes from it* its 'ideal entities'" (p. 224, my translation and emphasis). That is, the ideal entities are "made from" "the givens of factual humanity" and those of the social and material "human Umwelt." It is out of concrete doing that ideal entities *are made* rather than existing in some abstract and abstracting mind in an ethereal, metaphysical world.

We may now return, to the critics of embodied nature of mathematical cognition. What we would research, therefore, are not Dubinsky's claims about the two expressions $\lim_{x \to \infty}\left(\frac{1}{x}\right) = 0$ and $\lim_{x \to 0}\left(\frac{1}{x}\right) = \infty$, the former of which he classifies as an equation of two numbers and the latter of which a convention. Dubinsky here states a claim about the nature of the two inscriptions, whereas our ethnomethodologically informed researcher would be studying the proof work itself from which Dubinsky wants to detach the two objects. In fact, to make his claims, Dubinsky entirely depends on the reliving of the experience of the equality and the distinctions he makes, a distinction that he suggests Lakoff and Núñez do not make in their work. But as soon as Dubinsky were to prove his point, in and through his embodied|social demonstration, and as soon as Lakoff and Núñez in and through their own embodied|social work reproduce the proof in and through their own embodied | social work, the objects can indeed begin an apparently independent life of their own.

The entire problem, therefore, can be reduced to the fact that the debate has been about mathematical objects as Galilean objects, that is, objects independent of their particular frame of reference, which is the lived work of mathematical proof. The critics of the embodiment approach talk about mathematical objects as (independent) Galilean objects and about a Kantian mind, whereas others, including the embodiment theorists, focus on the initial work from which mathematical objects emerge. But the Galilean reduction inherently cannot have access to the work of doing mathematics, because it puts out of play the senses and sensible world (Henry, 2000). The very approach of Núñez's critics has disqualified any notion of the role of the body in cognition. Neither approach therefore really attends to the interdependence of the two aspects, namely the fact that mathematical objects exist only in and as of their role in mathematical, always *embodied* activity. This interdependence constitutes the very possibility of mathematics as cultural-historical (inter/subjective) and transcendental (objective) phenomenon (Husserl, 1939): "In the connection of mutual (participative) understanding, the production and result of one subject are actively understood again by others" (p. 212, my translation). Whereas the reproduction and transformation of geometry in practical engagement is necessary to keep it alive (rather than as a "dead language," such as Latin), this reproduction does not give it the objectivity. It is precisely the written proof account that allows the ideal mathematical object, which is virtually in the world, to be reproduced as such at any instant in time.

Critics such as Dubinsky might claim that some mathematical objects are purely "mental constructions" not related to anything in the world. But the point is, as Husserl (1939) shows that the sciences (including geometry and mathematics) have built idealities upon idealities beginning with entirely embodied concepts and by means of, as Livingston (1986) shows, socially accountable, entirely embodied mathematical work. The higher order idealities do not just exist, but they exist in and through the embodied | social practices that reproduce and transform the classical proofs and their extensions to new domains. The embodied nature does not disappear with more complex or "abstract" mathematical entities "because the logic of science produces from expressions with sedimented referential content only further expressions of the same character" (Husserl, 1939, p. 216, my translation). That is, the embodied | social nature of mathematics does not disappear with more advanced or abstract expressions (content); rather, it is precisely that the *character* remains the same in and with the advances that a science makes. We have shown how one can think of mathematical concepts in a new way so that concrete embodied experiences and words, as singular plurals, come to denote the concept as a whole, as a plural singular (Roth & Thom, 2009a). New opportunities therefore present themselves to study simultaneously heretofore held separate moments of mathematical cognition and learning. An exciting era lies ahead of us.

NOTE

1. The four grey rectangular triangles in the left square are the same as those in the right square. Thus, the area of the white square in the left square (the square over the hypotenuse) is equal to the area of the two white squares in the right square (the squares over the two sides). Anyone requiring some hint: label the sides of the triangle a and b and its hypotenuse c, then show that there the four triangles are equal (congruous). It can then be shown that $c^2 = a^2 + b^2$ because of the equality of the grey areas in the left and right squares.

REFERENCES

Bakhtin, M. M. (1986). *Speech genres and other late essays*. Austin: University of Texas Press.

Bakhtine, M./Volochinov, V. N. (1977). *Le marxisme et la philsophie du langage* [Marxism and the philosophy of language]. Paris: Éditions du Minuit.

Derrida, J. (1989). *Edmund Husserl's origin of geometry: An introduction*. Lincoln: University of Nebraska Press.

Dubinsky, E. (1999). Book review: Mathematical reasoning: Analogies, metaphors, and images. *Notices of the AMS, 46*, 555–559.

Garfinkel, H. (1967). *Studies in ethnomethodology*. Englewood Cliffs, NJ: Prentice-Hall.

Goldin, G. A. (2001). Counting on the metaphorical. *Nature, 413*, 18–19.

Henry, M. (2000). *Incarnation: Vers une philosophie de la chair* [Incarnation: Toward a philosophy of the flesh]. Paris: Seuil.

Husserl, E. (1939). Die Frage nach dem Ursprung der Geometrie als intentional-historisches Problem [The question about the origin of geometry as intentional-historical problem]. *Revue internationale de philosophie, 1*, 203–225.

Lakoff, G., & Núñez, R. (2000). *Where mathematics comes from: How the embodied mind brings mathematics into being*. New York: Basic Books.

Livingston, E. (1986). *The ethnomethodological foundations of mathematics*. London: Routledge & Kegan Paul.

Livingston, E. (1987). *Making sense of ethnomethodology*. London: Routledge & Kegan Paul.

Lynch, M. (1997). *Scientific practice and ordinary action: Ethnomethodology and social studies of science*. Cambridge, England: Cambridge University Press.

Rorty, R. (1989). *Contingency, irony, and solidarity*. Cambridge, England: Cambridge University Press.

Roth, W.-M., & Thom, J. (2009a). Bodily experience and mathematical conceptions: From classical views to a phenomenological reconceptualization. *Educational Studies in Mathematics, 70,* 175–189.

Roth, W.-M., & Thom, J. (2009b). The emergence of 3d geometry from children's (teacher-guided) classification tasks. *Journal of the Learning Sciences, 18,* 45–99.

Vygotsky, L. S. (1978). *Mind in society: The development of higher psychological processes.* Cambridge, MA: Harvard University Press.

ABOUT THE AUTHORS

Paul Cobb is a professor of mathematics education at Vanderbilt University, Nashville, Tennessee.

Brent Davis is professor and David Robitaille Chair of Mathematics Education at the University of British Columbia, Vancouver, British Columbia.

Laurie D. Edwards is a professor of education at Saint Mary's College of California, Moraga, California.

Charles Hohensee is a graduate student at San Diego State University, San Diego, California.

Götz Krummheuer is a professor of mathematics education in the Department of Informatics and Mathematics, Institute for Mathematics Education and Informatics Education at the Johann Wolfgang Goethe University, Frankfurt am Main, Germany.

Jean-François Maheux is a PhD student in the Department of Curriculum and Instruction at the University of Victoria, Victoria, British Columbia.

Ricardo Nemirovsky is a professor in the Department of Mathematics and Statistics, San Diego State University, and is director of the Center for Research in Mathematics and Science Education, San Diego, California.

Rafael Núñez is a professor in the cognitive science department at the University of California at San Diego, California.

Luis Radford is a professor of mathematics education at Laurentian University, Sudbury, Ontario, Canada.

Chris Rasmussen is an associate professor of mathematics education in the Department of Mathematics and Statistics at San Diego State University, San Diego, California.

Wolff-Michael Roth is the Lansdowne Professor of Applied Cognitive Science at the University of Victoria, Victoria, British Columbia.

Jennifer Thom is an assistant professor of mathematics education in the Department of Curriculum and Instruction at the University of Victoria, Victoria, British Columbia.

Carrie Tzou is an assistant professor of education at the University of Washington at Bothell, Washington.

Megan Wawro is a graduate student in the Center for Research in Mathematics and Science Education at San Diego State University, San Diego, California.

Ian Whitacre is a graduate student in the Center for Research in Mathematics and Science Education at San Diego State University, San Diego, California.

Michelle Zandieh is an associate professor of mathematics education at Arizona State University, Tempe, Arizona.